Workers of the Earth

‘In a time of planetary crisis, ecofeminist political ecologist Stefania Barca offers a brilliant analysis of how the diverse strands of twenty-first-century left politics can deepen their mutual understanding and move forward in a unified struggle for Life on Earth.’

—Ariel Salleh, author of *Ecofeminism as Politics*

‘A vital contribution to making sense of the planetary proletariat and socialism in the web of life.’

—Jason W. Moore, author of *Capitalism in the Web of Life*

‘Stefania Barca’s innovative reconceptualisation of the relationship between labour and the environment provides an indispensable theoretical basis for creating a better place for all living beings on this planet.’

—Kohei Saito, author of *Slow Down: The Degrowth Manifesto*

‘Stefania Barca has given us the book that agents of a socially and ecologically just transformation need, connecting women workers in paid and unpaid care work, subsistence workers, indigenous workers, industrial workers, workers caring for human and non-human beings in her insightful analyses that allows environmental labour studies to overcome the compartmentalisation of labour.’

—Nora Räthzel, Umeå University

Workers of the Earth

Labour, Ecology and Reproduction in the Age of Climate Change

Stefania Barca

First published 2024 by Pluto Press
New Wing, Somerset House, Strand, London WC2R 1LA
and Pluto Press, Inc.
1930 Village Center Circle, 3-834, Las Vegas, NV 89134

www.plutobooks.com

British Library Cataloguing in Publication Data
A catalogue record for this book is available from the British Library

ISBN 978 0 7453 4387 7 Paperback
ISBN 978 0 7453 4391 4 PDF
ISBN 978 0 7453 4390 7 EPUB

This book is printed on paper suitable for recycling and made from fully managed and sustained forest sources. Logging, pulping and manufacturing processes are expected to conform to the environmental standards of the country of origin.

Typeset by Stanford DTP Services, Northampton, England

Simultaneously printed in the United Kingdom and United States of America

Contents

To Marco, for everything we share

Acknowledgements

This book would not have come to light without the support and inspiration from many people and places over the past two decades. First of all, the movements, organisations and networks with which I have interacted most over the past few years: the Global Women's Strike – especially its London-based Women's Crossroads Centre, and the Global Climate Jobs Campaign – especially its Lisbon-based branch Climáximo; the Comitato Cittadini e Lavoratori Liberi e Pensanti of Taranto; the Just Transition Research Collaborative at the United Nations Research Institute on Social Development (UNRISD); the Just Transition and Care network, and all the people who have contributed to it since 2021.

My special thanks go to Pluto's editor David Shulman, for the immense patience, encouragement, competence and care which he has gifted me with; to all those who have offered help, advice and inspiration for the writing of this book, and especially Catia Gregoratti, Rocio Hiraldo, Selma James, Emanuele Leonardi, Nina López, Giulia Malavasi, Mario Pansera, Gea Piccardi, Rosa Porcu, Maurizio Portaluri, Nora Räthzel, Virginia Rondinelli, Chris Sellers, Dimitris Stevis, Irina Velicu, Francisco Venes; and to some who are not with us anymore, but will never be forgotten: Angus Wright, Peter Taylor and Peter Waterman.

Last but not least, I wish to thank the Zennström Climate Change Leadership programme at the University of Uppsala, the University of Santiago de Compostela, the Interuniversity Research Center for Landscapes and Cultures (CISPAC), the Xunta de Galicia, and the European Commission Horizon 2020 grant 101003491 (Just2CE project) for offering me the time and resources to investigate and contribute to the development of a feminist Just Transition.

Acknowledgements

Introduction: Labour in the Great Acceleration (1945 to Present)

Labour and working class are concepts rarely found in environmental research and discourse. But does this mean that they are of no relevance to our understanding of today's planetary crisis? Based on research in environmental history and political ecology, this book offers an unusual narrative of environmental change, one in which labour matters. Work and environment have long been construed as opposing realities; however, little is known about their relationship, historically and at present. This book contributes to expanding our understanding of this relationship, by looking at how both waged and unwaged labour – in industrial, domestic and subsistence work – as well as their organisations and movements have been experiencing environmental and climate change, and how they have been acting with respect to it. The answers offered here are not straightforward and might sound surprising or unlikely at times. In a sense, this is what I hope to convey: a feeling of displacement, which might turn into new ways of seeing things, and expand our political imagination.

For the most part, the book engages with the labour/ecology nexus over the last six decades, a period which earth-system scientists have termed the Great Acceleration (GA), characterised by the unprecedented degradation of earth systems due to exponential economic growth on the global level. Based on a collaboration between natural science and historical research, the GA era was first defined in 2004 and updated in 2015, and it remains to date a highly influential historical account of the planetary crisis.[1] Its aim was 'to capture the holistic, comprehensive and interlinked nature' of the post-1950 changes in humanity/earth relationships.[2] To do so, twelve socioeconomic indicators were chosen to represent 'the major features of contemporary society': population, GDP, international finance, urbanisation, energy, fertiliser and water use, large dams, paper production, transport, telecommunications, and international tourism. These were correlated with twelve indicators of earth-system change: the atmospheric concentration of carbon dioxide, nitrous oxide, and methane; stratospheric ozone loss; surface temperature; ocean acidification; marine fish capture and shrimp aquaculture; nitrogen flux into coastal zones; tropical forest loss; domesticated land and terrestrial biosphere degradation (see Figure 1).

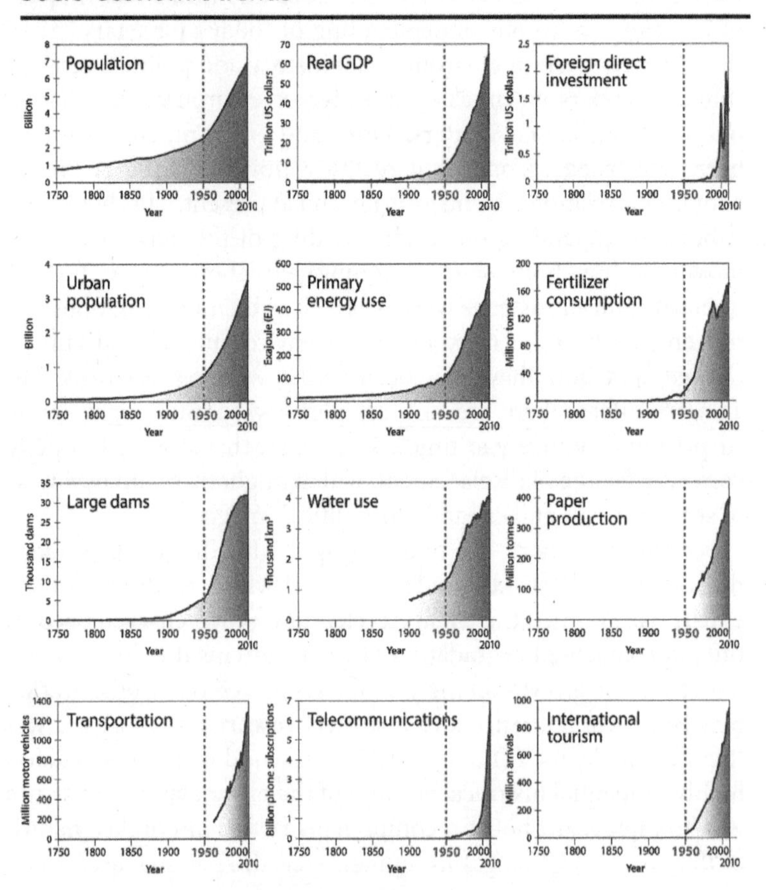

Figure 1a The Great Acceleration graphics
Source: International Geosphere Biosphere Program (IGBP). The Great Acceleration data can be downloaded from: www.igbp.net/news/pressreleases/pressreleases/planetarydashboardshowsgreataccelerationinhumanactivitysince1950.5.950c2fa1495db7081eb42.html (accessed 16 September 2023).

The Great Acceleration is a well-documented, yet incomplete picture of the last six decades. While useful in evidencing the earth-threatening capacity of (some) industrial technologies, global trade and finance, it does not adequately represent the workers of the world: though workers, like all human beings, are clearly part of nature, their experience of environmental and climate change is not accounted for in the GA narrative. The GA graphs represent humanity as the master of earth, in the act of expanding its ecological footprint by growing in numbers, and by extracting, consuming, and wasting resources from the biosphere; they do not represent humanity as a living part of the earth, made

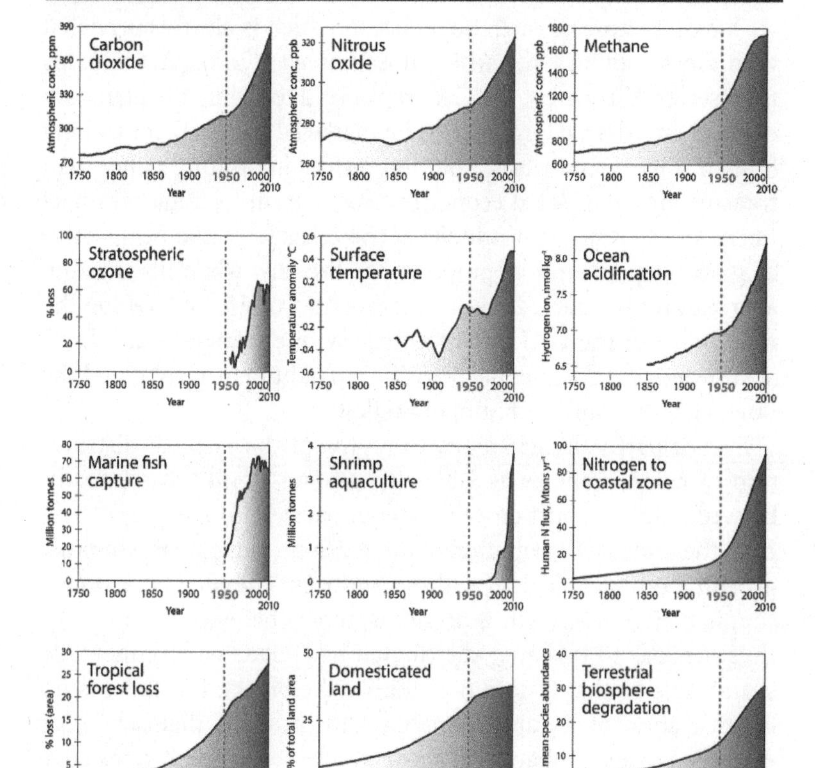

Figure 1b The Great Acceleration graphics

of and impacted by the same elements that make earth systems. In other words, the GA graphs do not show us human vulnerability: we miss the flesh and blood of human bodies, made of the same substance with the earth and other creatures, and threatened by the same processes that also threaten the biosphere. We remain with the impression that the growth of global Gross Domestic Product (GDP) has made humanity invulnerable, master of the earth, and that only non-human nature is threatened by it. In fact, the distinction between the two sets of data which are correlated in the GA graphs reinforces dualistic understandings of humanity versus nature which are themselves part of the problem. In Figure 1(a), we have a picture of activities which – although unevenly – are seen as unquestionably beneficial to humans; in Figure 1(b), we have a picture of earth degradation which is clearly correlated with those activities, but which does not seem to impact humans themselves. Ultimately, the GA graphs lead to seeing the planetary crisis as an external 'problem' to be managed by an all-controlling human subject, and whose solution resides in limiting rather than transforming the global economy. This, I argue, is a master's narrative, in the sense that it looks at the crisis from the perspective of global capital, and of power more broadly, while discounting as irrelevant the perspective of those who work for it.[3] Telling the right story of the GA, one in which labour matters, starts from acknowledging the unequal distribution of human agency, and the unequal vulnerability of human bodies.[4]

Even if partly disaggregated to represent the unequal distribution of responsibilities and benefits among world areas (mostly, between OECD and non-OECD countries),[5] the GA graphs still omit the unequal distribution of the *social costs* of global economic growth, both globally and within countries. The implicit assumption is that GDP growth is an undisputable benefit for humanity, thus it needs to be better distributed and made environmentally sustainable so that it does not destroy the planet. But there is no consideration of its unsustainability for humans themselves, or else, the adverse impact that those accelerated growth rates have had upon human life and labour; for example: the exponential increase in occupational accidents and long-term illnesses due to industrial toxicity; the loss of access to safe food and water caused by petrochemical or radiation exposure upon a number of ecosystems; or the mass evictions and forced migrations caused by large-dam constructions and other development megaprojects – all of them disproportionately affecting working-class, peasant and Indigenous communities worldwide.[6]

The lack of representation of human depletion and exhaustion as strictly related to the depletion and exhaustion of non-human nature is neither casual nor inconsequential. It originates from biases which are cultural and pre-scientific in nature – prejudgements that inevitably orientate the choice of data, their correlation and their organisation into a coherent narrative. On the one hand, the GA graphs reflect the hegemonic common sense of the age of growth, which sees commodity production, energy consumption, international trade and the associated GDP indicators as equivalents of human wellbeing, discounting their social costs, and devaluing all that is seen as non-productive 'activity'. At the same time, this picture reflects the neoliberal fantasy according to which labour does not matter – nor does class. In fact, the socioeconomic charts of the GA lack any recording of occupational trends, wages, working conditions in industry, agriculture or the service sector; of labour relations and access to the means of production; of the changes in the organisation of work. The accelerated growth of production over the past six decades appears as a natural phenomenon, something which happened by its own law of motion, rather than the result of human labour. As a consequence, people only appear as consumers whose needs and desires have been fulfilled to the highest point by the capitalist/industrial mode of production. Not as the ones who have been sacrificed to it, whose bodies and minds have been put to its service, or who have resisted it in different ways. In this fantasy, workers are one with their masters, parts of a whole which is called 'humanity' (or OECD countries at best), standing opposite to earth. And yet, most chapters of this book will tell you stories in which workers are one with the earth – threatened by the advancement of master's control over nature (both human and non-human), vulnerable to environmental degradation, and struggling against both – showing how labour and class matter in making sense of the GA era.

An even more implicit assumption of hegemonic common sense, reflected by the GA graphs, is the patriarchal fantasy that reproduction does not matter. Although the socioeconomic trends of the GA graphs start with population growth, no data is shown in the charts about the tremendous overburdening of reproductive and caring labour which has been implicated with the sevenfold growth in the number of humans along the past century. Population growth is taken as an indicator of human progress and wellbeing, and correlated with resource extraction, consumption and waste, but its social cost is discounted as irrelevant. In 2018, right before the Covid-19 pandemic struck, the ILO[7] estimated that 45 per cent of all working hours performed weekly at the global level were spent in unpaid care work, and that women and girls did around 75 per cent of it. This tells us that: (1) almost half of all human labour is not accounted for by the GDP measure of wealth; (2) this devalued labour is gendered; and (3) most reproductive work is freely appropriated by the global economy to sustain its growth, primarily via the provision of a cheap labour force.[8] None of this is visible in the GA charts. If industrial/waged labour does not matter, even less does the mostly unwaged labour of working-class and peasant women who have been forced into the accelerated production of life that falls into the category of 'population growth'.

Reproductive work is strictly associated with environmental and climate change. In both rural and urban areas, domestic labour, and especially parenting and nursing the young and the sick, is heavily affected by environmental conditions and hazards, including radiation or electromagnetic exposure, POP contamination and toxic waste, water and air pollution and other problems; not to mention catastrophic climate events.[9] As ecofeminists and environmental justice activists have long demonstrated, all these hazards tend to concentrate in the bodies and territories of the less affluent and most discriminated against, with the inevitable result of intensifying the burden of care work in these communities. At the same time, reproductive work is associated with care for the biophysical environment. Since the mid-1980s, studies have shown that much of the unpaid work done by rural women and girls was devoted to provisioning food via subsistence farming, which implied taking care of soil, water, seeds, plants, barn animals, and regenerating the biophysical conditions for both human and non-human life.[10] In peasant, fishing, herding and Indigenous communities all over the world, subsistence provisioning is largely dependent on the preservation of healthy local ecosystems, and is threatened by the continuous expansion of extraction frontiers and by the 'grabbing' of land, water, oceans, and anything in them, on the part of corporate and state actors, in the pursuit of GDP growth. The clash between growth and subsistence-oriented economies has been connected to the global rise in environmental conflicts over the whole GA era, and to the central role played by rural women in them.[11] In short, reproductive work is associated with the work of caring for earth and environmental reproduction, as well as to environmental protest and action – the two key dimensions of what ecofeminists have called *earthcare*.[12]

And yet, in the patriarchal fantasy of human wealth, labour is associated with the work of producing commodities, but not with that of producing life; with the perspective of exchange value but not with that of subsistence and use value.[13] Imagined as one with master humanity – industrial technologies, global trade, commodities – labour is implicitly held responsible for the earth degradation consequent to productive developments, but not for people- and earth-caring as related to subsistence and reproductive work. Yet this book will tell you that both domestic and environmental reproduction work have made a difference in the GA era, showing how sex/gender and race/coloniality also matter in making sense of it. In the Epilogue, I will draw some considerations on the significance of reproductive work for a labour-centred post-carbon transition.

Overall, the book will show how these two perspectives – that of commodity production and that of life-making – are blurred in the actual lives, bodies and agency of most workers, precisely because they are living beings and not machines. Both perspectives are thus fundamental to understanding labour's position in the hegemonic system of capitalist/industrial modernity. By this expression, I mean a specific type of modernity – that which considers the forces of production (Western science and industrial technology) as the key driver of human progress and wellbeing, while considering reproduction (both human and non-human) as a passive instrument to industrial production, and to the infinite expansion of GDP. This paradigm sees both the earth and care work as necessary resources, to be appropriated and maintained as cheaply and as efficiently as possible.[14]

A RESEARCH AND POLITICAL AGENDA

This book was compelled by an urge to shed light on the ecology of labour in the contemporary era, including labour's vulnerability to earth degradation, its agency in resisting the extraction, consumption and wasting of non-human nature, and its earth-caring and life-making capacity. The fundamental premise is that the dramatic environmental changes of the industrial age have directly affected workers in a variety of ways, and this has turned them into ecological subjects. The book builds upon environmental history and political ecology research, to which it contributes from a feminist ecosocialist perspective.

Environmental historians have offered a fresh perspective on the social and cultural drivers of the work/nature conflict, illuminating histories of environmental protection and conservation as performed by different kinds of workers, often with the support of trade unions. As shown in **Chapter 1**, *Labouring the Earth*, Environmental history has helped identify labour and working-class environmentalism as a distinct form of environmental consciousness, fundamentally different from wilderness conservation as performed by governments or markets. Building on this body of scholarship, and inspired by British socialist scholar Raymond Williams, the chapter connects stories of workers in different environments, both rural and urban, across Europe and the Americas in the nineteenth and twentieth centuries, reading the labour/ecology nexus through three different lenses: the landscape, the community and workers' organising. It focuses on physical work performed under distinct labour regimes: for example, reclaiming Italy's wetlands from malaria, extracting rubber and other commodities from the Amazon forest, coal from Colorado and oil from Northern Mexico, cultivating oranges in California. These labours are seen through the lens of workers' bodies, their embeddedness within specific power relations (of waged as well as enslaved labour, of class and racial discrimination), the energy that is extracted from them to reshape different environments into value-production, and their vulnerability to dangerous environmental conditions. To this, the chapter adds the perspective of the working-class communities, tracing the place of labour in environmental justice mobilisations, and the perspectives of different kinds of labour organising – from large trade union confederations in Italy and the United States to rural unions in Brazil – and the important difference that some of them had made to environmental politics between the 1960s and the 1990s.

'Labouring the Earth' results from a decade of investigations into the historiography, but also literary, oral and visual sources, that explored the labour/environment connection. My interest in this was not purely academic, however. I was trying to make sense of how working-class people were positioned in the ecology of industrial capitalism, because I believed this would help expand the political imagination of both ecology and labour movements. The chapter concludes by suggesting that labour-friendly sustainability policies can only emerge from acknowledging how the labour/environment dichotomy is a cultural and political construct of the industrial era.[15]

The nexus between industrial labour and ecology, as manifested through occupational and public health risk, had become clear to me for the first time a decade earlier, while watching a documentary on Italian TV dedicated to workers suffering from serious pathologies related to the uncontrolled development of organic chemistry. This included the story of Nicola Lovecchio (1947–1997), a worker from Manfredonia, Apulia, who had initiated a large workers' mobilisation for recognition and compensation of the serious health damage they had suffered from their job. I thus discovered that a chemical accident had occurred at Manfredonia's Anic plant (later to become Enichem) in September of 1976, only a few months after the Seveso accident (July 1976), with similarly disastrous consequences for the local community and territory.[16] Though I found no accounts of this disaster in scholarly narratives of Italy's postwar era, a wealth of sources was in fact available – including workers' oral testimonies, and those of their widows, theatre pieces, and much more. I found out that, almost unknown to the Italian public, this story had reached the European Court of Human Rights, after a decade of mobilisation on the part of Manfredonia's women, particularly those from the working-class neighbourhood of Monticchio, demanding an adequate recognition of the damage that the company had caused to their families and domestic life. In short, Manfredonia's story seemed to me like a microcosm where a whole new history could be written of the age of growth, and of labour's position in it.

Starting from there, I sketched a research agenda that followed the links among occupational and public health, militant research, trade unions, industrial labour, domestic labour, and environmental justice trials in working-class communities north and south of the country. **Chapter 2**, *Bread and Poison*, a comparative analysis of struggles against industrial toxicity in Italy between the 1960s and 1990s, offers an early account of what I came to understand as 'labour environmentalism', that is, a specific form of environmental mobilisation, which, unlike the ecology movement typically described in environmental history and sociology research, started from and revolved around labour. Not only the waged labour of industrial workers, literally struggling to breathe throughout the new hazards of organic chemistry, and seeking alliances with militant doctors and other scientists via trade unions, but also the unwaged labour of working-class women, struggling for a recognition of the value of working-class lives, and of their work in producing them.

The chapter recounts how, already in the late 1960s, a vision of ecology as having to do with the industrial manipulation of nature via human labour became the theoretical basis for trade unions' engagement in the regulation of hazard in the country's so-called industrial triangle. This political move had important repercussions upon the emergence of Italy's environmental movement in the same period, leading to the development of a working-class version of ecology. The chapter contrasts this with the events later occurring in Manfredonia, which showed how the environmentalism of Italian trade unions, born out of Italy's 'economic miracle', and focusing on male blue-collar workers in industrial jobs, was not considering the colonial and sexual divisions of labour characterising Italian capitalism of the time, reflected in workers' experience of industrial hazards. Nevertheless, a different type of working-class environmentalism had manifested in the south, and women – as unwaged domestic workers – had played a leading role in it.

While the environmental agency of organised labour is starting to be widely acknowledged in the specialised literature, most attention has been given to waged workers' relationship with the environment through trade union's bargaining, strikes and political lobbying; little if any attention has been paid to unwaged labour as a substantial part of anti-mastery ecological agency.[17] An historical example of the original contribution given by working-class women to environmental politics is examined in depth in **Chapter 3**, *Refusing 'Nuclear Housework'*, which focuses upon mobilisations against the nuclear industry in the UK in the late 1980s. The chapter tells the story of a group of women who organised, as part of a wider movement, to oppose the projected construction of a new reactor at Hinkley Point, in Somerset. They were part of one of the most well-known political organisations of second-wave feminism: the Wages for Housework (hereafter, WFH) campaign, which was born from women's struggle for financial independence, on the basis of recognition for the reproductive work they were already doing. In the UK, mothers had won family allowances and single mothers could claim a benefit from the state, while in the United States, a mass movement led by Black mothers on welfare was campaigning for a guaranteed income. First launched in Manchester in March 1972, this demand was embraced by autonomous Marxist feminists in Italy; the international WFH campaign which formed as a result launched a path-breaking challenge to the politics of women's emancipation by claiming that all women were contributing to society and the capitalist system with their unpaid domestic work, and demanding that this work be recognised as such and remunerated.[18] The WFH campaign has been widely criticised because its insistence on the necessity of remunerating housework – which the movement now describes more fully as reproductive and caring work – was at odds with the way in which both liberal and Marxist feminists understood emancipation;[19] moreover, two of its Italian spokespersons, Silvia Federici and Mariarosa Dalla Costa, had already left it by the end of the 1970s.

Nevertheless, the WFH campaign has continued for five decades since then in the form of a composite international and intersectional movement, whose core vision has been maintained through changing contexts and political alliances, and has achieved some important results in terms of both global and national policies.[20] Further, the WFH movement, which now coordinates a Global Women's Strike movement, has recently joined forces with the global climate justice movement by demanding that governments in various countries institute a Care Income for people of all genders who do unpaid care work in homes, communities, the land[21] and the environment; this is meant to end people's dependence on carbon-intensive jobs and to reorient green policies towards global social justice rather than green profits.[22] This recent development, together with the previous involvement of the WFH campaign with anti-nuclear activism, make it a very interesting case for political ecology research, and particularly for better understanding the nexus between labour and ecology as mediated by (unwaged) care work.

Unfortunately, a comprehensive history of this movement is still to be written. Even when recognising its foundational role in the domestic labour debate, most scholarly accounts of the campaign discard it after a quick read through some key texts from the 1970s. Very rarely do we find a serious engagement with the writings of the movement's long-term leader, Selma James, a non-academic writer whose intellectual work spans five decades of sex-race-and-class international activism;[23] or from key contributors and founders of the movement's Black and Caribbean branches,[24] or of the several autonomous organisations which are part of the campaign in various parts of the world. As a result, few feminists today know about this movement's history, involving not just the stereotypical white housewife but women of colour, disabled women, sex workers, women farmers, migrant women, paid domestic and care workers, trans women, and men (especially anti-war activists), organising against women's poverty as central to their opposition to sexism, gender-based violence, racism, ableism, militarism, colonialism, and exploitation in all its forms.[25] I believe a comprehensive analysis would show how these autonomous but interrelated struggles constituted the essence of the movement's strategy for working-class unity, based on a substantial redefinition of the working class to include the unwaged of the world – while never excluding the waged.

The WH campaign, so often reduced to a demand, is an international perspective to reclaim the centrality of caring as a fundamental social relation, and of carers as a political subject, whose labour is necessary to both human and capital's reproduction – hence, as embodying a key capitalist contradiction, that of capital versus life.[26] As long as reproductive work remained invisible and unpaid for, the movement claimed, the mainly women doing it were in a weaker position to make their collective voice heard, thus their potential for challenging the system was hindered. This, I believe, made the WFH a labour movement in its own terms, a sort of social movement unionism, whose strategy consisted in claiming adequate compensation for reproductive labour as a means to limiting capital's power over reproductive workers, by expanding their material wealth, physical and mental wellbeing, autonomy and bargaining power.

Over the past five decades, this approach to caring has constituted the core concern of the WFH movement, and its original contribution to anti-master struggle worldwide.[27] Without pretending to do full justice to this movement's history (which would require a whole book), Chapter 3 will start by revisiting it from the unusual but revealing perspective of its engagement with environmental activism, by organising Black and other working-class mothers against nuclear power.

This understanding of labour environmentalism is then expanded in **Chapter 4**, *Taking Care of the Amazon*, towards including the labour of environmental care. Commonly understood as part and parcel of the 'forces of production', labour is also, at the same time, a reproductive capacity, expended not only in producing the human species and society, but also the biophysical conditions for its wellbeing. The work of environmental reproduction – caring for, restoring and protecting soil, water and biodiversity (or else, the earth commons) – is a constituent part of what I call the 'forces of reproduction', that is, a historical subject, whose material and political agency is key to labour environmentalism. Throughout the industrial age, I argue in Chapter 4, this subject has counteracted the species supremacy that is the constituent of the capitalist project by enacting what I call 'interspecies commoning', that is, anti-mastery relations both among humans and between them and non-human nature.

This argument is informed by the life stories of Maria do Espirito Santo and Zé Cláudio Ribeiro da Silva, two earth defenders from the Brazilian Amazon, assassinated in 2011, who had achieved international resonance and gained a posthumous Heroes of the Forest award from the United Nations Environment Programme (UNEP). As they both lived and worked in one of Brazil's agroforestry areas denominated 'extractive reserves', instituted after a decade of struggles from a coalition of Indigenous organisations and rural unions called the Alliance of Forest Peoples, I realised that Zé Cláudio and Maria were not simply victims, but members of a collective subject whose history was missing from Anthropocene narratives.

Being a typical locus of modern environmental discourse, forests are probably the place where labour has been hidden the most and most successfully in both cultural and political representations. Nevertheless, they are also places where the ecological contradictions of capitalist/industrial modernity have been profoundly enmeshed, and where key historical turns have taken place in the relationship between labour and the environment. This is certainly the case of Maria and Zé Cláudio. Chapter 4 will enter their story via an analysis of oral and visual testimonies of their lives and work, speculating on its significance for advancing our understanding of the labour/environment nexus, as well as for the political project of labour's environmental leadership.

Taken together, the first four chapters of this book make clear how labour environmentalism must be understood as something much broader than its blue-collar version, because both domestic and subsistence workers, or else reproductive labour, have played significant agency in the history of environmentalism. Yet, as the remaining three chapters will argue, this recognition contrasts with the political ecology of labour as developed over the past five decades – that is, from the first oil shock of 1973 to present-day 'climate jobs' and Just Transition discourse.[28] The idea of Just Transition (JT) emerged in the 1970s from trade unions in the United

States as a means to alleviate the discursive tension between the protection of jobs and environmental protection, demanding measures that would protect the industrial workers affected by environmental regulations. In the 1990s, Canadian and US labour activists used the JT concept as part of their mobilisations, and in 1997, they brought it to the Kyoto climate change negotiations. In 2006, the Argentinian government endorsed the concept as part of its negotiation strategy at the United Nations Framework Convention on Climate change (UNFCCC). In 2010, the International Trade Unions Confederation (ITUC) made a resolution to combat climate change through sustainable development and just transitions. Since then, the political influence of the JT concept has grown rapidly, becoming integrated in the programmes and strategies by governments, bilateral and multilateral development agencies. For example, in 2015 the International Labour Organization (ILO) launched a set of 'guidelines for a just transition'. Similarly, the European Union (EU) has recently created a JT fund. Furthermore, at COP 27 in Sharm El Sheik (2022), the ILO launched the first ever Just Transition COP Pavillion, to discuss JT measures and commitments at the global level, including: a 49-country initiative on Climate Action for Jobs, a new Green Jobs for Youth Pact, and the launch of products that support the financing of a Just Transition and nature-based solutions.[29]

Chapter 5, *Greening the Job*, offers a critical appraisal of the ITUC and ILO formulations of JT, as merged and consolidated between 2010 and 2015, that is, as investments in low-emission and labour-intensive technologies and sectors, coupled with training programmes and income support for newly unemployed workers in polluting sectors. In this formulation, the post-carbon transition appears as a more sustainable capitalism that can come about through 'dialogue and democratic consultation' with 'social partners and stakeholders', alongside 'local analysis and economic diversification plans in order to help local governments to manage the transition to a low carbon economy and enable green growth'. These stakeholders are seen to have a larger role to play beyond mere consultation: governments pass economic stimulus measures; corporations implement social responsibility policies; academics and political leaders advocate 'ecological modernisation' legislation; international organisations issue directives, reports and recommendations.[30] The ILO and ITUC were keen on emphasising the economic benefits of JT as well. Echoing the landmark Stern Review (which was released in 2006 at the behest of the British government and argued that the economic costs of fighting climate change were far less than the costs of inaction), the ITUC declared that mitigation action actually aids employment. Central to their approach was the notion that government intervention can balance costs and distribute benefits among social parties. If carefully planned, they argued, infrastructure for mitigation and adaptation can make climate change a job creator. One major problem with this apparently win-win strategy, I argue, consisted in overlooking the likely impact of massive infrastructure projects on local communities and ecosystems, and the green labelling of many traditional forms of production – like cash-crop farming for biofuels – now employed to inflate the green economy ledger.

Overall, Chapter 5 argues that, while the JT strategy seemed like good news in terms of demonstrating labour's capacity for planning a post-carbon transition, unwaged and reproductive labour – that is, more than half of the global working class – remained left out of it, thus seriously limiting its political potentialities. Trade unions and workers are charting a new course in the long history of labour environmentalism, one in which green growth and a Just Transition promise the economic growth and security that the Fordist dream once held out. But buying into this new dream, I argue, will not save organised labour from the shortcomings and constraints that have all but destroyed its strength in most countries. In fact, if they continue supporting capital's 'green' restructuring of the global economy, trade unions will find themselves on the opposite side of peasant and Indigenous communities, landless rural workers, unpaid domestic and social reproduction workers, subsistence farmers, and all those who bear the costs of 'green' capitalism – fostering renewed cycles of dispossession and subjugation.

Seeing this contrast as a political dilemma, what I called labour's eco-modernist dilemma, **Chapter 6**, *Labour and the Ecological Crisis*, offers a historical analysis of what I have identified as its origins in Marxist thought. Focusing on the work of four Marxist intellectuals whose ideas resonated with various social movements across the Left spectrum (labour, environmentalism, feminism and degrowth), the chapter describes four different but also interlacing trajectories of labour environmentalism in Western Europe over the last quarter of the twentieth century: the 'ecology of class' approach, as theorised by Italian communist writer and politician Laura Conti; the 'liberation from work' approach, as inspired by the Austro-French writer André Gorz, a key source of the degrowth idea; the eco-socialist approach as envisioned by the British socialist Raymond Williams; and the subsistence perspective proposed by German ecofeminist Maria Mies. These historical trajectories suggest that the current entrenchment of labour within the politics of eco-modernisation hides a number of internal fractures and alternative visions of ecology that need to be spelled out in order to open the terrain for a rethinking of ecological politics in class terms today.

This argument is then mirrored by **Chapter 7**, *The Labour(s) of Degrowth*, which criticises a certain understanding of degrowth as a political perspective in which labour plays no active part. Written as a contribution to a dialogue between ecosocialism and degrowth theorists, the chapter starts by noticing that the degrowth debate so far has lacked a clear vision of what social subjects, and which processes of political subjectivation, can turn its vision into a political strategy. Making space for labour in the politics of socio-ecological revolution, I argue, means aiming for a truly democratic, workers-controlled production system, where alienation is actively countered by a collective reappropriation of the products of labour and by a truly democratic decision-making process over the use of the surplus. Such strategy must be based on an extended concept of class relations that goes beyond the wage labour relation, and towards a broader conception of work as a (gendered and racialised) mediator of social metabolism. I conclude that ecosocialist degrowth should take the form of a struggle for dealienating both industrial and meta-industrial labour.

The book concludes with an **Epilogue** which reflects on my current political engagement with ecosocialist, degrowth and ecofeminist movements. Focusing on two international campaigns – the Global Climate Jobs and the Care Income – and on the international Just Transition and Care network, this chapter will offer some insights on the still undeveloped unity between waged and unwaged, industrial and meta-industrial labour, which is necessary in facing environmental injustice, climate change and extinction in the present.

PART I

History

1

Labouring the Earth: Transnational Reflections on the Environmental History of Work*

In July 2012, an Italian court ordered the forced closure of the Ilva steel plant in Taranto, the largest and one of the oldest such factories in Europe, finding it guilty of serious violations of environmental regulation and of causing what the court recognised as an environmental and public health disaster. Immediately, the case spurred wide attention from the national media, reporting on a long series of citizens' protests against the decision.[1] *There is no such thing as a safe environment/economy relationship* – Taranto's protesters seemed to argue. *Working-class people know better, as occupational and environmental risk have been part and parcel of their own history for centuries. They have made a living out of daily negotiations with all sorts of hazards and with death itself. The idea of sustainability is a typically bourgeois pretension, an illusion invented to obscure the reality of economic development in the industrial era. So, why and in the name of what – they seemed to ask – should our entire mode of life and work be destroyed? What will replace it? How can this new reality be worth our sacrifice?*

And yet, not all workers opposed the court's decision. A number of them instead protested against the company, the government and the municipality's absurd decades-long silence in response to obvious environmental and public health hazards. Together with a number of other citizens' organisations, these workers claim that

---

* This chapter is a slightly revised version of Stefania Barca, 'Laboring the Earth. Transnational Reflections on the Environmental History of Work', *Environmental History* 19, no. 1 (2014): 3–27. Reprinted with permission from Oxford University Press.

working-class people have a right to breathe clean air, drink clean water and live in a safe environment. They firmly contest the false dichotomy between health and work, silently implying the sacrifice of the environment, and call attention instead to the criminal responsibilities of the company in disregarding environmental regulations.

Taranto's dilemma exemplifies a crucial issue involving the definition of a possible transition to a more sustainable economy in a way that could be acceptable to, even actively shaped by, workers and working-class communities. Sustainability policies aimed at social justice must be based on new and convincing forms of articulation between labour and environmental issues. Environmental history can help in this endeavour by offering an informed and reflective view of the ways in which the work/environment conflict has been historically shaped and of how this conflict has in turn shaped the environment and people's lives.

This chapter offers a tentative framework for such analysis by pointing to three arenas where the connections between work and environment can be investigated. The first presents the landscape as reflective of past human labour. The second examines the workplace and its relationship with the local community. The third focuses on working-class and labour environmental activism. The discussion builds on research developed by both environmental historians and other scholars, referring to different scholarly traditions and focusing mostly on three contexts: Italy, Latin America (with special attention to Brazil) and the United States. By bringing these disparate literatures into a transnational dialogue, I aim to transcend the confines of national histories and historiographies in the hope that these encounters will cross-fertilise the research field and prompt new conversations on places and cultural contexts beyond those discussed here.

SEEING LABOUR THROUGH NATURE

Historical materialism is an excellent starting point for such reflection. Marx and Engels viewed labour as part of nature, in that they saw workers as natural beings exercising their physical as well as mental abilities on external nature. Labour and nature thus constituted each other in a dialectical, metabolic relationship.[2] The alienation of humans from labour was part and parcel of their alienation from nature, a product of enclosure and dispossession (so-called primitive accumulation) within the capitalist system. Marx's critique of capitalism is consistent with an ecological critique: capital organised the exploitation of nature (the natural conditions of production) through the exploitation of human work – a view that passed on to the Frankfurt School and that has also been shared by many environmental historians.[3]

This aspect of Marxian thought has influenced scholars interested in developing a holistic approach to society–nature relationships. In his 1980 essay 'Ideas of Nature', British literary critic Raymond Williams criticised the triumphalist view of the Enlightenment and industrial eras that asserted the separation of humans from the non-human world. Williams argued instead that human and non-human nature are inextricably linked through the labour process, stating, 'We have mixed our labour with the earth, our forces with its forces too deeply to be able to draw back and separate either out.'[4] More recently, ecomarxist scholar James O'Connor suggested that labour's role in the history of nature is that of a partner in a common story of domination and exploitation. What nature and labour share, he argued, is their being treated as commodities by the capitalist system. O'Connor invited scholars to see the history of labour as an important component of environmental history, noting that 'the more that (human-modified) nature is seen as the history of labor, property, exploitation, and social struggle, the greater will be the chances of a sustainable, equitable, and socially just future'.[5]

Similar views have been adopted in a number of environmental history narratives. A notable example is Richard White's 1995 essay 'Are You an Environmentalist or Do You Work for a Living?' There he noted that two ideas were deeply rooted within US environmentalism: first, that work is the main cause of environmental destruction, and, second, that an Edenic relationship (of whites) with the American environment had been possible at the time of first contact and was still possible to recapture. White argued that both ideas were rooted in Judeo-Christian culture that viewed labour as originating in sin and coinciding with the expulsion from Eden, where nature is a garden and a place of leisure, to earth, where nature is degraded into a wilderness and a place for sweat and fatigue. Contrary to this vision, White contended that work – not leisure – is the experience historically most significant to the human understanding of nature and that American environmentalists need to see nature as a place for human work and living. His study on the transformation of the Columbia River, The Organic Machine, applied a view of environmental change as a co-production of humans and nature's work.[6]

Along similar lines, Italian scholar Piero Bevilacqua views nature as an active historical agent, cooperating with labour in the creation of wealth. His 1996 book *Tra Natura e Storia* (Between Nature and History) presented an ecological critique of classical economic thought, including Marxian political economy, centred on a labour theory of value, which he stated was reducing nature to a passive object of manipulation. Such thinking, Bevilacqua argued, negatively affects our ability to see nature as an agent other than society and with its own economy, and thus prevents a true understanding of the interaction between the two. In his history of modern Italy, Bevilacqua consistently describes nature as a partner with labour in co-constituting landscapes and social formations. *Tra Natura e Storia* describes various configurations that the work–nature relationship had taken in the nineteenth-century Italian landscape, each corresponding to different land property assets and agrarian labour relationships.[7] In so doing, Bevilacqua builds on a tradition of Italian scholarship that started with Emilio Sereni's 1961 history of the Italian agrarian landscape, *Storia del paesaggio agrario*, a work that enjoyed international resonance in its field.

Conceived as a counterpart to contemporary histories of the French countryside by Annales scholars and based on visual representations (primarily paintings and maps), Sereni's book convincingly showed how the evolution of the Italian landscape since pre-Roman times was the result of the dynamic interaction of environmental conditions (climate, soil, altitudes, etc.) with different forms of socially organised work. In Sereni's account, generations of Italians had reworked the landscape they had inherited, with its resources and constraints, not only according to the economic, political, technical and cultural conditions of the time, but also in accordance with a consciousness of the place, that is, the aesthetic perception and the knowledge that people gained of the place by living and working on the land. In short, Sereni had adopted a view of the landscape as 'past human labor' incorporated into the soil.[8]

Although both Sereni and Bevilacqua can be considered Marxian historians, such a view of the agrarian landscape is not peculiarly Marxian. In fact, the perception of Italy as an 'artificial homeland' – a land made inhabitable through work – and of its landscape as an 'immense repository of human labor', was formulated in the mid-nineteenth century by the agronomist Carlo Cattaneo, based on the historical experience of Lombardy's Po River plain and in dialogue with contemporary observations by the English agronomist Arthur Young.[9] Nowhere was such a view more convincing than in the case of *bonifica agraria*, the drainage-and-improvement schemes repeatedly pursued since the early modern era. The *bonifica* thoroughly reconfigured the peninsula – particularly the Po plain, Tuscany's lowlands, the Pontine marshes and large parts of the south – between the nineteenth and twentieth centuries. Before (and even partly after) the introduction of the steam-powered pump, draining the land had required long, painstaking manual work using buckets, shovels and spades, with the help of horses and mules. Thousands of labourers from throughout the region worked in highly unhealthy conditions, waist-deep in water for most of the time. Many were destined to get ill and even perish from malaria, typhoid fever, or merely fatigue and malnourishment. The history of the peninsula itself, its geomorphology, is thus profoundly enmeshed with that of the people who gave the land a new shape by opposing their bodily force to that of the current, struggling against gravity and erosion, and finally reworking the course of its rivers.[10]

What must be recalled, however, is that the *bonifica* had military overtones. The conquest to reclaim land from water entailed harsh discipline and exploited profoundly unequal social relationships. Although performed under different social configurations and political regimes (from enlightened reformism to agrarian capitalism, fascist rule and postwar development politics), the *bonifica* always rested on one crucial element of the Italian agrarian landscape: the abundance of cheap labour, whether local or forcibly relocated from other regions. *Bonifica* was thus synonymous with the power of redesigning nature by means of controlling labour.[11]

Like the Italian *bonifica* landscape, but on a much greater scale, California's modern landscape reflects hard work performed under conditions of social domination, as both environmental and social historians have shown. Don Mitchell suggested that California's twentieth-century landscape is inseparable from the oversupply of itinerant labour, which was socially produced to make agriculture a highly profitable enterprise.[12] Likewise, Douglas Sackman described California's Orange Empire as a form of 'hegemony over people and places', obtained not only by recruiting its labour force from across the globe but also by extending its sphere of influence through advertising. Its promotional language created a vision of the citrus industry 'as a spontaneous production of Eden, bearing no traces of workmanship' – and especially of its working and living conditions.[13]

The Edenic language can be used not only to conceal, but also to sublimate the human work embedded in landscapes. In *No país das Amazonas* (In the Country of the Amazons, 1921), Portuguese director Silvino Santos portrayed the wonders of Amazonia as a frontier garden rich in every kind of natural wealth and diversity. He did so not through the language of aesthetic contemplation, but that of work. The film displays an incessant movement of men, women, animals and instruments in the act of working and reworking the place. The connotation of such work was largely positive. It was the ability of humans (both whites and native peoples, although in different ways) to alternatively make use of and contrast the forces of nature, typically exemplified by the currents of the Amazon River, in order to extract its resources and carve a space for themselves in the region. The film celebrated human work at the same time and by the same means through which it celebrated nature. Of course, Santos came after almost four centuries of European colonisation of the region; thus, what he filmed was not a 'native' environment.[14] Nevertheless, he demonstrated that even a quasi-wild and diverse landscape such as the Amazon could be celebrated as a co-product of human and nature's work.[15]

No páis das Amazonas provides us with an organic appreciation of the work–nature relationship, one that seems to contemplate the possibility of harmonious cooperation, of co-production. It gives an image of nature and human work as complementing each other: nature is not an untamed wilderness, even though its forces, beauty and wealth are still unspoiled; work is not a destructive force, even though humans are able to penetrate the jungle and get their hands on the incredible richness that comes from it. Perhaps such vision is one typical of frontier landscapes in an early stage of exploitation (when nature still appears as the ruler), and it serves to encourage human settlement and the business that goes with it. But visions such as these are destined to endure, even when the landscapes and social relationships they represent have long gone, as it is with the rubber industry that set the stage for large parts of Santos' movie and generously funded his enterprise. They express an idea of harmony that overcomes classical dualisms of Western culture between Eden and earth, of nature as leisure versus nature as sweat and fatigue, and so they function as what Carolyn Merchant calls 'recovery narratives'.[16] *No país das Amazonas* represents a recovery narrative of the Amazon in times of early frontier capitalism.

Of course, a good environmental history narrative can only take this as a starting point. Clearly, celebrations such as that of Santos' movie served to conceal the unequal and exploitative labour relationships that governed the remaking of Amazonian landscapes, and of course what came to be celebrated in the film is not nature per se but the resources it yielded to human ingenuity and hard work. Furthermore, the work that features in the movie as a co-protagonist with nature is that of racially discriminated migrant workers, debt peons, and enslaved Indigenous people. Blatant violations of human rights, mass deportations and slaughter were not uncommon – and they are still reported in some cases today – as a means to coerce local labour, particularly that of Indigenous populations who better knew the nature of the place, into working for capitalist enterprises, both national and foreign, in the tropical environment.[17] It is worth remembering that roughly 30,000 Putumayo people were reported to have died while working in rubber extraction for the British-Peruvian Amazon Company of Julio Cezar Araña in the first decade of the century.[18] *No país das Amazonas* was profoundly enmeshed in those power relationships. In fact, Araña funded Santos' professional film training and then hired him to produce a document that would capture Araña's version of the story – one where labour and nature were celebrated instead of annihilated.[19] And yet the film does represent human work as a co-constitutive element of the natural landscape, and in so doing it gives us the opportunity to go beyond the misconceiving and silencing of its own narrative to wonder what it truly meant to live and work in the early twentieth-century Amazon.

Viewing landscapes as products of past human labour is a first, crucial step for an environmental history of work. Raymond Williams noted how 'A considerable part of what we call natural landscape ... is the product of human design and human labour, and in admiring it as natural it matters very much whether we suppress that fact of labour or acknowledge it.'[20] The extent to which labour is part of the landscape varies with geography as much as with history. In any case, removing the image of work from the landscape has produced the dominant vision of nature in the Western/industrial culture that has led to romantic views of 'the environment' as something to be protected from work and, therefore, from working-class people, even though environmental historians and political ecologists have documented that private property, the market and state control, not labour, have been the main forces behind resource exhaustion.[21]

Romanticising people's perceptions of and relationships to nature through work is also problematic, as Thomas Andrews' study of the 'workscape' of coal mining in Colorado exemplifies. As with the *bonifica* works of the Pontine marshes and elsewhere, Colorado's coal workers experienced a 'cartography of risk', punctuated with explosive gases, invisible coal dust and falling rocks. That such danger was a result of complex combinations of social and natural forces – the political economy of fossil fuels and of corporate mining, in this case – on which individual workers had little control, also formed a crucial common perception for Colorado colliers who developed intense and radical forms of political mobilisation in the attempt to turn that perception into revolution.[22]

A similar story of political consciousness and revolutionary action arising from the experience of working in dangerous extractive activities is that told by Myrna Santiago on the oil fields of northern Veracruz, where a violent remaking of environmental, working and living conditions produced what she called the 'ecology of oil'. According to Santiago, the ecology of oil had come out of three interrelated processes, all together reconfiguring the relationships between nature, labour and social power in the local landscape: first, the dispossession of indigenous communities and of local settlers-farmers; second, massive environmental alterations, including forest and river habitat destruction, in order to make space for oil extraction operations; and third, the imposition of new, racially discriminating labour relationships. This triad of socioecological change was accomplished by the new liberal and pro-business elite of the late nineteenth century, who shared with foreign investors an ideology of progress based on the mastering of both labour and nature. For Mexican manual workers and their families, who occupied the bottom of the labour hierarchy, working and living in the Huasteca oil fields became a daily struggle against the hazards of the weather, tropical diseases, fire, and chemical and bacterial contamination. Lacking any substantial protection, they suffered a heavy toll for the so-called development of the Mexican oil industry. As a result, Santiago notes, the landscape of Huasteca's oil fields is literally lined with the corpses of racially discriminated local workers buried alongside pipelines.[23]

But this is not the end of the story, because the same oil fields turned Mexican workers into politically conscious and active citizens in the form of 'strikers, troublemakers, risk takers and union men'.[24] Mexican oil workers realised that property – that is, control over both the land and the labour process – mattered in determining the sustainability of their work experience in the tropical forest. In fact, Santiago argues, the 1938 nationalisation of the oil industry can be considered partly a result of three decades of workers' struggles for healthier and safer work conditions. Unfortunately, they came to learn that states do not always perform better than private industry in terms of environmental safety and human health. Even though the publicly owned PEMEX made notable improvements, including recognition of a number of occupational diseases, prevention training and medical treatments, production targets still remained more important than the preservation of workers' bodies or of the forest ecosystem. Moreover, the scaling up of oil operations implied the extension of oil risk from mostly workers to entire communities. Thus, not much changed in terms of labour–nature relationships, insofar as oil operations still rested on the destruction of the ecosystem and the impairment of living conditions for human and non-human life in the region.

In sum, the work–nature relationship that landscapes embody is crucially mediated by social relations. Be it in Italy's drainage improvement schemes, in California's agrarian landscape, in Colorado coalfields or the tropical 'frontier', controlling labour has created new socionatural orders that are profoundly unhealthy for both the land and the people. Adaptation and cooperation with nature have been replaced by attempts at domination and control, with often unwanted, unforeseen and irreversible consequences. It is not work per se, therefore, but the social meaning and ends of work that make the difference in the sustainability of society–nature relationships.

THE ECOLOGY OF WORKPLACE AND COMMUNITY

Another important step in the environmental history of work comes from placing the workplace centre stage in our narratives and understanding it as an ecological system. This approach was suggested by Arthur McEvoy in a 1995 article, noting that, from the vantage point of the shop floor, 'Ecology points to an analysis of health and safety in terms of the interaction between a number of systems: the worker's body and its maintenance, the productive processes that draw on the worker's energy, and the law and ideology that guide them.' Not only the workplace, but also workers' bodies should thus fully enter environmental history narratives as metatexts where the political ecology of industrial societies had been written.[25]

An important contribution in that sense was given, again in the United States, by Christopher Sellers' *Hazards of the Job*, a work that marked a turn in the literature by bridging the history of the workplace with that of environmental science and environmentalism. The book showed how US and European workplaces had been important spaces for knowledge production about human and environmental health and for professional coalitions pushing towards regulation of industrial hazards. Not only has work been extracted from workers' bodies in the course of the industrial era, but so too has knowledge. The branch of medical science known as industrial hygiene developed out of extracting information from workers' bodies and observing their reaction to a variety of risk factors in the course of their work life. This kind of science evolved in Europe and the United States between the last decade of the nineteenth century and the first half of the twentieth, and it reached a wider significance for the environmental movement through Rachel Carson's *Silent Spring*, which amply relied on research from physicians and industrial hygienists. It was that science that first began to draw the boundaries between normality and abnormality, acceptable and unacceptable limits of exposure and contamination. The environmental movement of the 1960s, according to Sellers, started from the criteria and definitions central to industrial hygiene to attack pollution.[26]

Studies of different workplaces have depicted them as typically unhealthy environments, where illness is normalised and workers bear the marks of the 'treadmill of production'.[27] Here environmental historians have profitably met with social historians of medicine and with scholars in occupational health, broadening the spectrum of possibilities for interdisciplinary dialogue beyond that of the natural sciences.[28] Workplaces, however, are far from being a typical subject of environmental history narratives, and especially so when it comes to industrial plants, where environmental historians seem particularly reluctant to enter. Yet the ecological dimension of modern industry is written in the flows of toxins emanating from the workplace to the environment and the human body through air, water, the biogeochemical cycles and the food chain. Before encountering the living environment outside, industrial toxins meet workers' bodies, which represent the biological dimension of the industrial workplace. Historically, such encountering between microparticles and workers' bodies gave rise to important forms of political and ecological consciousness, and to organised action.

A case in point is the story of labour environmentalism in Italy. Here, a vision of ecology as having to do with the industrial manipulation of nature and human labour became the theoretical basis for a consistent wave of radical leftist environmentalism. In *What Is Ecology: Capital, Labor and the Environment*, first published in 1977, scientist Laura Conti posited organic chemistry and carcinogenic, mutagenic and reproductive risk at the centre of a plastic explanation of ecology as the interconnection of all living and non-living matter. Conti's definition of ecology was very similar to that of another scientist who convincingly argued that petrochemicals pose a terrible menace to all living creatures including humans: the American biologist Rachel Carson. Unlike Carson, however, Conti was also a politician. She was an elected councillor for the Communist Party in the Milan district between 1960 and 1970, then in the Lombardia regional government between 1970 and 1980, and a deputy in the national parliament from 1987 to 1992, where she worked at the Agriculture Commission.[29] She never disentangled her commitment to environmental issues from her political engagement; the two were linked in a unique vision of society–nature relationships. In fact, she was the first in Italy to define political ecology as 'the study of how social relationships within the human species influence the natural world and other species'.[30]

Conti, however, was not alone in her search for ecological Marxism in Italy. In the autumn of 1971, at its yearly cadres' school in Frattocchie, the Italian Communist Party held its first national meeting on the theme 'Man, nature, society', emphasising the need for the party to consider the environment a working-class priority.[31] This opened the possibility for a left-wing environmentalism to take form in the country. Legambiente, today a highly established environmental organisation, arose in 1979 as a subsection of the Communist Party's cultural and recreational activities, and Conti figured among its founding members.[32] The rise of this Italian working-class environmentalism should be understood within the context of two decades of massive industrialisation and environmental transformation, at the end of which the country's public opinion (not only the leftist one) was ready to acknowledge the existence of an 'environment' problem.[33]

The event that marked the birth of a new ecological consciousness in the Italian Left was the Seveso accident. On 10 July 1976, the explosion of a chemical reactor at the ICMESA plant located near the town of Seveso, in Lombardy, caused a cloud of dioxin to rise over the town and its rural hinterland, directly affecting a population of 10,000. Of all industrial disasters during the Italian economic boom, the one in Seveso spurred the greatest attention on the part of the government and the national and international media, leading the European parliament in 1982 to pass the first European Union law on industrial hazards, known as the Seveso Directive. As a regional councillor, Laura Conti found herself at the forefront of the battle for citizens' 'right to know' and participative science that characterised the political relevance of the accident.[34]

Conti's battle for public access to information and decision making in Seveso revealed one interesting aspect of the political definition of risk: in explaining how the maximum acceptable concentration (MAC) of dioxin had been established, government officials declared they had relied on 'US standards for farm work', a statement for which Conti could find no evidence despite great efforts.[35] Along with other Italian labour physicians, Conti accepted the idea of using MAC levels of dioxin in occupational health science as a basis for delimiting Seveso's 'unsafe zone' and especially for defining the cleanup operations. Unfortunately, this was not the position adopted by the Italian authorities, either at the local or the national level. And, moreover, this was not what local people wanted. Whether they were misinformed or genuinely convinced that the risk was worth taking, Seveso residents wanted to stay home as much as possible. More, they longed for reassuring answers that spared them the painful choice of staying or leaving, having children or aborting. They clearly manifested their attachment to the place where they lived and worked, even though many of them were only first-generation migrants to that place. The Seveso experience added to Conti's vision of ecology a sense of the role of culture and symbolic meaning – places and people's connection to them must find their way into the science of ecology, she concluded.[36]

Seveso was a laboratory experience for what urban ecologist and communist militant Virginio Bettini termed 'class ecology', an approach theorised in the course of community meetings featuring left-wing scientists (Barry Commoner among them) that the communists organised in Seveso with the intent of mobilising local people against corporate and government coverup. The 'class ecology' approach was centred on industrial pollution as the most compelling and politically relevant aspect of the environmental crisis and on working-class people as its primary victims. Bettini claimed, 'Society's debt towards nature is a debt towards the working class.'[37] It is not clear, however, how far the working class, and even the workers of the ICMESA plant, actively participated in Seveso's 'popular scientific committees' organised by the communists, and on what positions. Despite their generous efforts at helping local people to struggle for their rights (and not only for monetary compensation), leftist activists in Seveso met with diffidence and even open resistance, also significantly related to their pro-abortion stance.[38]

The problem with the 'class ecology' approach was that, however ideologically correct, it met with the unexpected opposition of working-class people. As Laura Conti came to realise,

> People had never been put in the condition to understand that, to have a healthy environment, it is necessary to sacrifice something: everything has always been done to get more salary, more cars, more highways, even – in the best cases – more hospitals and schools, but almost nothing to get cleaner air, cleaner water, safer food. At this point, why expect that all of a sudden the Brianzoli [the people from Brianza, i.e., the Seveso area] recognise that living in a healthy land is worth a mass exodus?

Conti directed her critique against her own party, which had never taken a real stance towards the protection of nature. She decried the stigmatisation of Seveso residents as 'immature' or 'stubborn' and concluded, 'None of us has the right to criticize the Brianzoli.'[39]

The story of Seveso holds many significant implications for environmental historians. First, it tells us that a good environmental history of work should take into account the workplace/community link. Not only in the sense of the national community or of broader communities such as those of parties and unions, but the local community, where people have faces and names, are family or neighbours, and where they share the same air, soil and water.

In other words, a good way of connecting work and environmental narratives is choosing working-class communities – including the workplace, workers, their families and neighbours, and the local landscape – as the subjects of our stories, for here it is that all the complexities and contradictions of the work–nature relationship come into play.

Second, working-class communities are far from being unified social entities entirely corresponding to theoretical definitions of their class identity, interests and behaviour. Nor does working-class environmental consciousness entirely coincide with the politics of labour parties or unions. Although the theoretical effort on the part of the Italian Left to make the environmental question a class issue may sound correct, the environmental consciousness of Seveso residents retained unexpected motivations, such as the attachment to the place and the desire to preserve the local landscape as it was – contaminated by some invisible chemical substance.

Other environmental history studies have made this point comparatively clear. In Chad Montrie's *To Save the Land and People*, Appalachian people opposed coal strip mining, outraged by the devastation of the land, the loss of resources such as soil, fish and game, the peril of landslides, and the massive alteration of the local landscape. However, Montrie also shows how some of the conflict took place between deep miners and surface miners, the first accusing the second of being outsiders who did not care for the land they were ruining. The United Mine Workers also took on an ambivalent position. It initially supported the regulation of strip mining because, unlike deep mining, it was a largely non-unionised activity. The union, however, was not a monolith. Different positions on the strip mining controversy existed among miners and union leaders and came out through internal conflict at different moments of the story. Labour was thus an active part in the strip mining controversy.[40]

The concept of 'workscape', as formulated by Thomas Andrews, might prove an innovative instrument for investigating working people's environmental consciousness. One important aspect of that concept, for example, is that of human–animal relationships. Be it the rats befriended by Colorado colliers as daily companions in their mining rooms and as living warning systems against imminent danger, or the mules they had as helpers and with whom they had to establish labour relationships based on a difficult mix of domination and cooperation, or the seabirds with which Hawaiian guano workers shared their own workscape, as shown by a recent article by Gregory Rosenthal, animals have played an important role in humans' relationship to nature through work.[41] In the Seveso accident, as well as in innumerable other cases of chemical or radioactive contamination, pets, barn animals, birds, and other creatures that share their living space with that of humans, are typically the first to present the effects of intoxication, thus signalling Rachel Carson's message that the fates of human and non-human beings are inextricably related through the sharing of common life-supporting systems. More than that, they are related through what ecocritic Stacy Alaimo calls transcorporeal bonds, made of material, cultural, sentimental and symbolic links among which labour relationships should also figure.[42]

Andrew Hurley's concept of environmental inequalities also helps to elucidate working-class connections to the environment. In his book on Gary, Indiana, Hurley offers a detailed and nuanced portrait of the many peculiarities of working-class experiences of nature and of environmental change.[43] Others have followed his lead, showing that working-class communities elaborate particular ways of dealing with the environment and with environmental change, mediated by their labour and by the social relations of production that take shape in the local space. These relationships generate unique forms of 'ecological consciousness' based on some combination of work experience and experience of place, mixed with beliefs, traditions and institutions of the culture of which the workers are a part. Additional factors that may affect workers' ecological consciousness are related to personal identity and life experience, such as gender, age, race/ethnicity, and family history, to elements such as personal skills and one's position in the local labour organisation and employment status, or whether one lives, or is born, in the place where he or she works.

Despite many contingent, internal stratifications and differentiations, however, working-class communities do share common experiences and often develop a strong sense of belonging and identity based on some form of control over the work process, its social meaning and its scope. They thus develop their own perception of the work/environment tradeoff that shapes their lives and the places in which they work and live. Their own bodies and mental capacities, as well as those of their families, are at stake in the continuous transformation of the local environment. They feel partially responsible for such environmental change, viewing it as a bargain they have to make in exchange for survival. Such bargains are often overly simplified as jobs versus nature, which obscures the nature and the diversity of environmental activisms that develop from working-class ecological consciousness.

WORKING-CLASS ENVIRONMENTALISM(S)

To find alternative answers to the false dilemma of jobs versus the environment as played out in Taranto and elsewhere, we should start examining the many occasions in which working-class movements have encountered environmental movements, generating various forms of 'labour environmentalism'. Although journeying through working-class or labour environmentalism is quite uncommon for environmental historians, there are places (both academic and physical) where this has been done successfully, especially in the last decade. Undoubtedly, one such place is the United States, where the picture of people's environmentalism, to use an expression from Chad Montrie's recent book, is now articulated and nuanced enough to give us much material for comparative reflections.[44] We will describe a few examples of labour environmentalism in Italy and in the Brazilian Amazon, highlighting differences and similarities that speak to the possibility of a true dialogue among scholars of the work–environment connection.

Popular environmentalism in the United States has a long history, as Richard Judd's 1997 work on the origins of US conservation demonstrates. Judd pointed to popular perceptions of nature by the 'common people' of New England as 'neither conservationist nor anticonservationist', as we define these terms today, but as 'a complicated mix of Christian theology, practical wisdom, economic incentive and second hand natural history'. Without idealising or romanticising such popular environmental culture, Judd's work, *Common Lands, Common People: The Origins of Con-*

servation in Northern New England, signalled the importance of understanding the contribution it had given to environmental law and policy, for 'to ignore this perception, to wave it aside in the battle to protect the environment ... is to court disaster'.[45] Similar insights can be found in Gunther Peck's 2006 article on labour in US environmental history, which posed the question of how the experience of the commons in nature – and that of alienation – had shaped the history of human work. Another important suggestion coming from the article was that of looking at the forms of nature's utopianism harnessed by the British and North American labour movements since the early modern period, in order 'to make radical critiques of capitalism, landlordism, and slavery'.[46] Lawrence Lipin's study of the 'progressive conservation' that came out of working-class environmental culture in early twentieth-century Oregon and Chad Montrie's *Making a Living: Work and Environment in the United States* further reveal ways that workers attempted to mitigate environmental decline.[47]

A number of authors also have addressed the relationship of labour and leftist culture with the new environmentalism of the postwar era, focusing on the contributions that unions and the labour movement in general have made to environmentalism. According to Scott Dewey, it was during the early postwar period that awareness of potential health risks from pollution 'became quite advanced in US working-class people in comparison to that of their fellow citizens'. During the 1950s, the United Auto Workers, through their president Walter Reuther and vice president Olga Madar, pressed the government for the regulation of gasoline emissions, even if this meant losing a number of jobs. In Madar's opinion, workers were first and foremost American citizens and 'neither they nor their children develop any immunity to automobile exhaust pollutants or any other'.[48] In the following two decades, oil, chemical, atomic, steel and farm workers' unions coalesced with some environmental organisations, leading to the passage of some major pieces of environmental regulations, such as the Clean Air Act of 1970 and the Clean Water Act of 1972.

In *Forcing the Spring*, Robert Gottlieb showed how the role of health professionals, coming from the ranks of the student, feminist, environmental and radical Left movements, was crucial in soliciting those reforms, supporting the labour movement in their implementation and pushing mainstream union organisations and their leaderships towards inclusion of environmental protection in their grievances. As a leader of the Oil, Chemical, and Atomic Workers International Union, for example, Anthony Mazzocchi was instrumental in the passage of the Clean Air and Clean Water Acts. Likewise, Cesar Chavez and the United Farm Workers first raised the issue of pesticide poisoning in the early 1960s as a unified struggle in defence of both workers' and consumers' health.[49] As Robert Gordon argued, between the late 1960s and early 1980s, 'workers, environmental activists, and union leaders across the country concluded that the spread of hazardous substances in the workplace and the spread of pollution in the environment represented two aspects of the same problem', and the cultural premises for organisational alliances between environmental and labour movements were laid out.[50]

The 1970s was a fruitful period for labour/environmental alliances in the United States, as exemplified in the experience of the Environmentalists for Full Employment groups, the Urban Environmental Conference, and Ralph Nader's and Barry Commoner's networks, among others. Nevertheless, according to Gottlieb, for environmental organisations, workplace and social justice issues remained external to their mission, just as the labour movement 'remained bound by union acceptance of the structure of industry decision making'.[51] Furthermore, as Brian Obach has argued, the relationship between labour and the environmental movements grew more conflictual during the 1980s, as a result of the political turnover of the Reagan era.[52] At the grassroots and local level, however, a number of cases have been documented in which coalitions continued well into the 1980s and 1990s, extending to the present.[53] At that point, working-class environmentalism was not just a matter of efforts at coalescing labour and environmental organisations, but it converged with grassroots and community struggles put forward by what came to be known as the environmental justice movement.

The environmental justice movement in the United States defines the environment as the place where 'we live, work and play'. In fact, as Robert Bullard pointed out in *Dumping in Dixie*, work has been a potent mechanism of environmental injustice and racism, considering that the most unhealthy low-paying jobs are those most likely to be filled by African Americans and Latinos. He wrote, 'Requiring people to choose between jobs or the environment is inherently unfair. The solution to this dilemma lies in making workplaces safe for workers. Anything short of this goal places workers at an unfair disadvantage.'[54] Largely reputed as the founding text for environmental justice studies and activism, *Dumping in Dixie* was built on a wide recognition of the importance of 'job blackmail' as a structural cause for the production of environmental injustice. Labour unions, however, rarely figure in the book, as the leadership for environmental justice activism had evidently shifted to different actors.

This move from union to community activism as the privileged terrain where grassroots environmental struggles are fleshed out has been interpreted in social science in terms of a shift from the conceptual framework of 'class' to that of 'subalternity'.[55] In the last decade, 'environmental conflict' has become an important way to describe subaltern environmentalism or so-called environmentalism of the poor, spurring a new array of social science research, also involving and intriguing environmental historians.[56] Environmental justice, subaltern environmentalism, environmentalism of the poor and environmental conflict are all useful concepts to understand various typologies of struggles coming from working-class people and involving environmental costs and benefits, both in the urban and in the rural space. These struggles often contain an unobserved or undertheorised link between labour and environmental concerns. Most social science research on environmental conflicts pays attention to community agency – sometimes assuming the term acritically – as opposed to government or corporate agency, while overlooking the role that workers play in such conflicts, or rather the relevance of work in mediating people's understanding of the environmental issues at stake. Paradoxically, work and its complex relationship to environmental concerns is probably the less known aspect of environmental justice struggles and of environmental conflicts.

And yet work is and has always been relevant to those struggles, for the simple reason that 'subaltern' people, racially discriminated people or 'the poor' are typically also working-class people, that is, people who occupy the lower ranks of the labour hierarchy, making a living out of the most dangerous and most unhealthy jobs while also living in the most polluted places. 'In every way', environmental sociologist David Pellow has written, 'the workplace is an environmental justice issue.'[57] In the United States as elsewhere, work has been a relevant component of environmental justice struggles since the beginning but in ways more complex and contradictory than expected. As Pellow's study of the garbage industry in Chicago demonstrated, for example, recycling programmes supported by environmentalists and even environmental justice organisations often turn into extremely hazardous jobs and labour-unfriendly policies, oppressive of workers' rights and culture. Anti-unionism, in fact, features as a main driver of the book's narrative; unions only figure as either suppressed or absent actors.[58]

Union activism, however, is not completely absent from environmental conflicts and justice, or from subaltern environmentalism. In some cases, in fact, union organisations and individual leaders have played a crucial role in defining the terms of the conflict and setting out the possibilities of change. To find what is probably the most compelling story of labour environmentalism, one that reached and still holds global relevance, we must once again turn to the Amazon region, particularly to the state of Acre, Brazil. Here, in the early 1980s, a landless workers' movement for the protection of the forest grew out of rubber tappers' struggles against powerful lumbering and ranching interests, and it led to the passage of an important piece of legislation, the 'extractive reserves' bill of 1990.[59] The 'extractive reserves' idea represents a very interesting possibility to overcome the work/nature dichotomy typical of Western culture. In fact, it is a good means of overcoming the West/non-West cultural dichotomy itself, for it is inspired, on the one hand, by Indigenous people's relationship with the land and, on the other hand, by the 1988 Brazilian Constitution that acknowledges and protects 'traditional' populations' rights to their ancestral lands. The extractive reserves idea put together the defence of different non-capitalist forms of the work–nature relationship: collective use rights (as opposed to individual property rights); land-based cultural identity and livelihoods (as opposed to forest encroachment and forced evictions); and wild fruit gathering and biodiversity conservation (as opposed to monoculture and resource exploitation). In all cases, the forest gets to be preserved not as a place for leisure or for scientific investigation, but as a place for working and living – indeed, as what allows the local human communities to survive and maintain their culture. Be they Indigenous or caboclos, the *extractivistas* (rubber tappers but also wild fruit gatherers and fishers) are those who actively protect the forest from encroachment and destruction; the forest needs those people as much as the people need the forest.

According to anthropologist Mary Allegretti, the rubber tappers' movement can be considered a new social movement, in the sense that it encompassed objectives of a different nature, at once social, economic, cultural and environmental, linked together by a coherent worldview and political project. Their actions span from primary education and adult literacy to biodiversity conservation, from access to livelihood resources to the protection of cultural identity, from health to transportation means. Taken together, all these things were aimed at allowing people to survive and even thrive in the forest environment. Moreover, the seringueiros adopted innovative forms of both organisation and struggle. The most typical was the empate (first occurred in 1976), an action in which men, women and children 'would stand unarmed in the way of tree cutters and their equipment, blocking the destruction and appealing personally to the peons as people of the same social class'. As Biorn Maybury-Lewis writes, 'It was a nonviolent, communitarian, educational, and consciousness-raising approach to struggle, where all involved on both sides went away thinking that "this is different, this is special."'[60]

However, like older social movements (and particularly those involving labour), the *seringueiros* had to confront the violence of police repression and assassinations of union activists. Among movement leaders who came to be prosecuted under national security law were Ignacio Lula da Silva (future president of Brazil, then leader of the metalworkers union) and Francisco 'Chico' Mendes, a local leader of the Sindicato dos Trabalhadores Rurais and founder of a national rubber tappers' council. Mendes was assassinated in 1988, and his death came to have a large international resonance that served to push the Brazilian government to pass the 'extractive reserves' project into law. The principle of nature conservation through sustainable use on the part of local populations was then recognised at the global level through the International Convention on Biodiversity of 1992. This represented the highest acknowledegment, for environmental policies, of the 'past human labour' that is embodied in landscapes and, at the same time, of the possibility that a more sustainable future rests on the human ability of 'saving' nature through work.

In other cases, however, labour unions have maintained a much more detached attitude towards environmental issues or even openly opposed grassroots environmental action at the local level. Still, this has not impeded workers' environmental activism. The Italian case is significant in this respect. Several important trial cases against large polluting companies, especially in the petrochemical and asbestos sectors, have been brought forward in the last 40 years stemming from occupational health grievances.[61] Typically, those struggles have been based on 'popular epidemiology' studies, collecting evidence about work-related cases of cancer among the labourers of those industries, but they have then become class actions – one might say 'working-class actions' – involving workers' families and larger communities, including the neighbourhoods around those factories, the urban population affected by air, water and soil pollution, local fishing, sports and environmental associations, women's organisations and health professionals. Italian unions have followed a historical pattern somewhat similar to that highlighted for the United States: they have strongly supported environmental regulation, especially in the industrial sector during the 1960s and early 1970s, while adopting a much more reductive approach to workers' grievances during the economic recession of the late 1970s. In some cases, local unions have even aggressively boycotted environmental justice actions, practising various forms of ostracism towards those members who supported them. Like the one taking place in Taranto today, those struggles have incorporated all the dilemmas and contradictions typical of the work–environment relationship in industrial societies, which makes them all the more interesting.

CONCLUSIONS

Like so many other working-class communities, Taranto's workers seem to perceive the environmental discourse as something alien to their world, not because they despise it (who is not an environmentalist, nowadays?), but simply because this is a choice that is not offered to them. Their democratic options and the exercise of their citizenship rights seem in fact to be limited by the position they occupy within the industrial order. In short, they do not seem to have a right to be environmentalists. Yet environmental histories of working-class people and movements seem to contradict common-sense assumptions about the work/environment conflict. Be it in the oil fields of the Huasteca or in the Amazon forest, in the Italian 'industrial triangle' or in the mountains of Appalachia, working-class communities, union leaders and individual workers have been historical subjects endowed with ecological consciousness and agency as advocates for environmental health and conservation. Their environmentalism, when and where this has taken an organised form, is simply different from that of the metropolitan middle class and of all those social groups whose subsistence does not depend on any specific agroindustrial activity.

Perhaps the answer to the work/environment dilemma may come from recent discourses of Just Transition, pointing at the injustice of shifting the cost of environmental cleanups on labour and invoking the right to compensation and the right to voice – that is, to self-determination on the part of working-class communities – as regards the shift to an environmentally sustainable economy.[62] In order to build 'Just Transition' policies, however, a first step is recognising the historical role of work as the single most important interface between society and nature, and of working-class people as possible subjects of a more inclusive vision of how to 'save the environment'.

2

Bread and Poison: Stories of Labour Environmentalism in Italy, 1968–98*

This chapter tells the story of the encounter between a generation of Italian experts in industrial hygiene (physicians and sociologists) and factory workers, and how that encounter translated into new forms of knowledge and political action. The chapter aims to highlight the material relations existing between occupational, environmental and public health as they were experienced by subaltern social groups, who knew industrial hazards through their bodies and through the environments where they worked and lived. This material reality – the organic relationship between humans and nature through work – has been politically obscured by dominant social forces and by the divide between the labour and environmental movements. The case of labour environmentalism in Italy, however, shows how, in particular places and at particular times, the possibility emerged for a reunified consciousness of industrial hazards, one that challenged alienating forms of scientific knowledge and political-economic power.[1]

The experience of labour environmentalism in Italy began when what I call 'militant science' – the new Italian industrial hygiene born out of the 1968 student movement – interacted with the knowledge of environmental hazards embodied by factory workers. During the 1970s, starting from a platform of health

---

* This chapter is a slight revision of Stefania Barca, 'Bread and Poison. The Story of Labour Environmentalism in Italy', in *Dangerous Trades. Histories of Industrial Hazards in a Globalizing World*, ed. Christopher Sellers and Joseph Melling (Philadelphia: Temple University Press, 2012), 126–39. Reprinted with permission from Temple University Press.

grievances based on a mix of scientific and lay expertise, the Italian labour movement drew up a comprehensive strategy of struggle for occupational, environmental and public health. That early coalition of labour with 'militant' industrial hygiene in Italy eventually produced legislative reforms of great social and environmental significance, such as the Labour Statute (1970) and the national public health system (1978).[2] Yet, as this chapter also shows, the actual impact of those institutional reforms on workers' bodies, work environments and local landscapes came to be biased by a history of political-economic differences. In dealing with workers and the environments around them, I aim to contribute to the building of an 'embodied' and environmentally conscious working-class history. I start by showing how, during the transformation of the country into a highly industrialised economy (1958–78), workers' bodies bore the marks not only of capital but of the industrial state, especially in the petrochemical sector. At the same time, I emphasise how, once they joined with 'militant' experts, those same bodies (and minds) openly challenged and counteracted dominant ways of knowing and regulating industrial hazards.[3]

THE ECONOMIC BOOM AND ITS SOCIAL COSTS

Between 1955 and 1970, roughly 4 million people from southern Italy migrated to the northern industrial regions, searching for factory jobs. From the 1960s onward, they found work in the fast-growing petrochemical, steel and mechanical industries. By the end of that decade, and during the 1970s, the Italian government implemented a new policy of transfer of industrial jobs to the south by locating a number of publicly controlled companies, mainly in the petrochemical sector, along the shores of the southern regions.[4]

In consequence of such massive changes, the country experienced the epidemiological shift typical of advanced industrial economies – namely, from infectious to degenerative diseases, especially those correlated with environmental poisoning from mercury and benzene hydrocarbons. Research in industrial hygiene began to be sponsored by the INAIL (the Workers' Compensation Authority) and by the European Community for Coal and Steel, but it was mainly focused on risk insurance and clinical pathology rather than on prevention in the work environment. Facing an impressive worsening of work conditions and a steady increase in occupational accidents, the Italian unions adopted a defensive strategy, mainly based on the attempt to increase compensation and strictly enforce it. Compensation law, however, was a major obstacle to the prevention of industrial hazards, for the same law sanctioned the total non-liability of employers in the matter of both workplace accidents and long-term health risks.[5]

EXPERTISE AND MILITANCY: THE 'ENVIRONMENTAL CLUB'

This typically market-oriented approach to workers' safety was to be abandoned and completely revised during the 1960s – a period in which union confederation was particularly strong politically – and gradually led to the passage of a very advanced piece of legislation, the Labour Statute of 1970. Coming after a decade of tremendous changes in the cultural and political climate of the country, the statute granted workers the right to exercise direct control over working conditions. This principle was revolutionary in the sense that it emancipated employees from the oppressive control of 'company doctors', whose behaviour was strongly conditioned by their being on the company payroll. Furthermore, the Labour Statute introduced a radically new conception of workers as assigners of physicians' services and thus entitled them to control over employers' choices. To enforce this right, workers were allowed to bring independent experts into the workplace to test its environmental conditions and to examine the workers' exposure to risks.[6]

Those 'experts', on whom workers relied for their empowerment in the workplace, were mostly young physicians and labour sociologists, coming from a student movement that in Italy was strongly hegemonised by the radical Left and considered itself – in Gramscian terms – an intellectual army at the service of the working class. During the 1968 university protests, students and researchers in industrial hygiene were invited by union representatives to collaborate with the union confederation in breaking a history of subordination of medical doctors to employers. 'Socialising Culture', the slogan of the student revolt in general, became particularly meaningful in the case of workplace injuries and diseases. Medical students neglected their university courses to study in the factories, learning from workers' testimonies.[7]

This was the golden age of the movement for workers' health in Italy. A permanent workshop, formed by sociologists and 'new' industrial hygienists under the political hegemony of the union confederation, elaborated a new scientific paradigm for the work environment based on the translation of complex analysis into a few simple principles of political action. These were embodied, mainly, in the slogan 'Health is not for sale' and in the principle of non delegation in the matter of health issues, implying the workers' direct control over knowledge and practices regarding the workplace. Soon renamed the 'Environmental Club', this group helped to redefine the new confederate political strategy for health and safety. Meanwhile, at its thirty-sixth congress held in 1972, the Italian Association of Industrial Hygiene officially recognised the 'objective' value of the workers' experience and the utility of a 'participative' methodology for the collection and recording of environmental and biostatistical evidence at the work group level. This was a methodology on which the Environmental Club had been working for a few years, based on the direct production of knowledge within the workplace through a series of practical measures that workers could carry out during their workday – such as monitoring noise, dust, temperature, and so forth.[8]

Between 1969 and 1972, unions' grievances directly regarding health and safety grew from 3 to 16 per cent. Most interesting, however, is that this struggle did not concern the work environment only: it was directed towards broader reform in national health policies, since unions and the Left parties were calling for a new system of public health services directly controlled by the state. A series of industrial accidents occurring during the 1970s, mainly in the petrochemical sector – and particularly the Seveso disaster of 1976 – were instrumental in keeping public opinion focused on the relationship of industrial hazards to environmental and public health. Risk prevention cancer epidemiology, exposure standards, right to know and participatory decision making became widespread ideas, leading to the formation of the organisation Medicina Democratica – a grassroots action/research movement that was to become instrumental in a number of occupational and environmental health controversies in the following decades.[9]

The importance of this particularly positive period of struggle and social alliances can be seen in its major accomplishment, the Public Health Reform Bill passed in 1978. This legislation mandated locally based public health services (USL) that would supervise both environmental and health quality within factories and communities. With it the principle of the internal relationship between workers' and citizens' rights to health obtained institutional recognition at the highest levels. The most important meaning of the workers' health struggle, therefore, was as a primary test for broader social reform, affecting the whole of Italian society. By struggling for a redefinition of pollution-related diseases, factory workers not only sought greater safety for themselves but also aimed for more comprehensive sanitary protection for their families and the entire national community. This story represents in some way the success of what unions, and the political Left in general, defined as 'the political strategy of class alliances and solidarity'.

The Italian public health system was born at the end of a long battle for occupational health and represented that battle's most significant victory. At the core of the fight was a new consciousness about the material and political unity of work, environment and public health – or labour environmentalism – that had first been tried on the shop floor.

HEALTH STRUGGLES NORTH AND SOUTH

The encounter between labour and environmentalism in Italy began at the heart of the country's most industrialised area, between Torino and Milano, in the core years of the Italian economic boom and in the middle of a revolutionary cultural transformation related to the student protests of 1968 through 1977. Where the joining of occupational with environmental and public health produced its most advanced results was the province of Milano in the period 1972 through 1977 – with the experience of the SMALs, which is the topic of the following paragraph. The chapter then moves through space and time towards a rural area of the south – Manfredonia, Apulia – where a very different scenario of labour environmentalism took form.

Milano: Reforming from Below

In 1972, the regional government of Lombardia instituted SMALs – Servizi di Medicina per gli Ambienti di Lavoro (Medical Services for the Work Environment), giving the service the task of supporting the implementation of the fifth and ninth articles of the Labour Statute, which concerned workers' right to control the enforcement of health and safety measures in the workplace. At the demand of the 'factory board' (a union representative committee), SMAL physicians entered the workplace to investigate the health conditions of the workers by measuring levels of hazard and compiling and updating 'environmental data' registers and personal sanitary journals for each worker. On the basis of their research, they made nosological inquiries in collaboration with public health agencies, and instructed employers regarding compulsory risk prevention measures. Most important, both the results of the SMAL physicians' research and the countermeasures they proposed were formally discussed with workers, through a 'health committee' overseen by the factory board. The SMALs were a public health service mandated by law with the purpose of integrating the work of existing public health services (namely, the Labour Inspectorate and INAIL), as these agencies had proved ineffective in halting the worsening of workers' health during the previous two decades. The creation of the SMALs gives us an idea of the extent to which the action/research methodology elaborated by the Environmental Club since the late 1960s had become culturally hegemonic and politically feasible. We might consider them a successful example of what science scholar Sandra Harding defined as the philosophy of 'strong objectivity' – that is, a research method that intentionally assumes the standpoint of victims and marginalised others. Supported by physicians and labour clinicians from the University of Milano and overseen by local and regional administrations, the

SMALs were granted authority based on scientific rationality, and unions could use their findings as a solid basis for labour disputes.

In fact, the SMALs were a form of 'militant' science. Their methods reversed the traditional industrial hygiene approach: now it was not workers as guinea pigs for occupational medicine, with scientists reading their bodies to extract 'scientific' data from them; it was the other way round, as workers themselves solicited the experts' intervention to give scientific support to their empirical observations about health hazards on the shop floor. But workers could only realise this in an organised way, through their factory boards. It was the factory boards – that is, the unions – that called for a SMAL intervention and finally decided what initiatives to carry out on the basis of the SMALs' recommendations. The control of the unions over all SMALs' activity is clear: they had pressed local administrations to create the SMALs and had lobbied for the passage of the regional bill; they organised courses and training activities for would-be SMAL physicians, selected candidates, and put them in contact with workers; and they set the SMALs' agenda and coordinated their activities at the regional level.[10]

It was not just a practical and political hegemony, however. The language of the SMALs' reports shows how physicians fundamentally shared with the unions a militant conception of knowledge as a form of empowerment, as well as a militant conception of health as part of the broader conflict between labour and capital. The SMALs interpreted their relationship with employers not in a defensive, but in a counteractive way. They entered the structure of production, starting at the plant level, and discussed the scope and regulation of technological change that is the very core of industrial production.

In the SMAL vision, in line with the insights of the 'new' industrial hygiene and the 'Environmental Club', technical progress and economic growth had produced in Italy, as in other advanced industrial countries, not a general improvement in health conditions but a shift in disease patterns. The types of pathology had changed, not their social incidence. This was true within the factory, where the classic distinction of risk factors (dust, chemicals and physical conditions) was to be aggravated by new factors, such as rhythm and position of work, standardisation, repetition and automation. Furthermore, given that most occasional illnesses tended to become chronic, 'the opinion that any health professional felt to give about the dangerousness of some work environment', according to a SMAL document, 'would be deficient and partial if not confronted with the opinion of those who live there eight hours a day'.[11] The workplace was to be seen as an (unnatural) environment, and workers were the ones who knew it best.

The SMALs' self-conception as militants is also shown by their behaviour as rank-and-file activists rather than as impartial, disinterested science professionals. In Cinisello Balsamo, for example, the local SMAL dealt with a complex social conflict, fostering local opposition to the Terzago steel plant because of a noise issue. It proved in this case to be much more than a health professionals' service, instead acting as an intermediary in an environmental conflict while also accomplishing its task of mandating stricter health and safety measures. SMAL physicians quickly connected the noise pollution issue in the community with the existence of a serious health hazard within the workplace, and acted to eliminate both at their source. That was not an easy task, however, because the situation was exacerbated by the factory owners' response to citizens' protests – forcing workers to close the plant's windows and thus aggravating the effect of the noise on workers' ears and the lack of ventilation in the plant. Furthermore, in this small-scale factory the union's presence was weak, so SMAL intervention had been called for by the local public health official responding to the demand by a citizens' anti-noise committee. The physicians' official entrance into the workplace as a bureaucratic agency could have upset the employees, who were worried about the employer's threat of shutting the plant down.

This case clearly shows the internal contradictions in the relations between workplace and environment and between labour and community in the matter of health. These contradictions led to a kind of intervention that was scientific and political at the same time, that was able to reconnect the two fundamental loci of the struggle (within and beyond the factory gates), and that was able to act at different material and political scales. Reassuring the workers and looking at ways of eradicating the noise problem required the SMAL physicians to adopt a militant vision of their institutional and professional task: it required them to accomplish tasks not strictly inherent in their mandate, such as setting up a series of community/workers' meetings with the participation of experts from the Otolaryngology Clinic of the University of Milano and members of the local government.

In their final report, the SMAL physicians diagnosed partial deafness in 30 per cent of the workers and chronic acoustic shock in another 36 per cent, mostly women. These results, based on 'objective' audiometric measurements and international standards, could not be denied by the employer. The SMAL intervention, though, did not stop at the workers' health conditions; the physicians sought the collaboration of 'democratic technicians and engineers', as they wrote in their report, meaning the voluntary support of external experts in solving the interrelated problems of vibration and transmission of acoustic waves. As a result, SMALs were finally able to suggest a whole variety of technical solutions for limiting noise and preventing future injuries, at the same time resolving the community/workplace conflict.[12]

Another case, that of Metal-Lamina, a metal mechanical plant in Assago, illustrates further the interconnections between the work environment, worker health and community health. In this case, too, the SMAL intervention had been demanded by public officials at the local level, starting with the Municipal Ecology Service, on the grounds of complaints coming from workers in a neighbouring plant about Metal-Lamina's discharges of smoke. The SMAL physicians found that the presence of lead dust within the workplace was ten times the legal standard and ordered the immediate hospitalisation of eight out of thirteen smelters. The workers reported that five dogs had died in the plant in the course of one year, probably by the ingestion of lead dust deposited on the ground.

The stopping of the foundry blocked production, and management threatened to shut down the plant; eventually, however, the company decided to implement all of the SMALs requirements and those of local public officials concerning the abatement of lead dust and smoke, and it installed a water purification system. The managers also asked the SMAL experts to become the company's consultants in the matter of health and safety regulations.[13]

This case opens up the question of managerial and entrepreneurial behaviour. The only reported cases of continued opposition to the SMALs' work are those involving the plants owned by Montedison, the most important partially state-owned chemical corporation in the country, producing synthetic fibres and pharmaceuticals in a number of plants. Montedison had merged with the ENI group (the State Agency for Hydrocarbons), forming Enichem, a powerful petrochemical company that came to own a number of plants for the production of fertilisers and plastics spread along the Italian coast. Its behaviour was representative of the particular contradictions that marked the Italian experience in the matter of worker and environmental health. In opposing the entry of SMAL experts into its plants, the Montedison management claimed the protection of workers' health as their exclusive business, accomplished by its medical service. The existence of such a company service and its partial control by the state were, in management's view, sufficient reasons to present the company's workers as a privileged category, which did not need supplementary oversight.[14]

The idea of the Montedison-Enichem workers as a privileged group was grounded in the state's involvement with the petrochemical sector, which was perceived as strategic production in the Italian economy. It was also the result of the power relations between unions and the state, which allowed Italian workers in public companies to have permanent, secure jobs. This complex mix of conditions gave the petrochemical industry in Italy immense social power, as we will see in the next case, concerning an Enichem plant in Manfredonia. Here the entrepreneurial state dealt with a rampant internal conflict of interests and functions, centred on the problem of risk definition and distribution of social costs.

Manfredonia: State Chemicals

The ENI group first arrived in Manfredonia – a fishing town on the Adriatic coast – in the late 1960s, under the name ANIC (Azienda Nazionale Idrogenazione Combustibili, State Company for the Hydrogenation of Carbons) to explore the methane layer in the area with the intention of building a plant for the production of urea and ammonium sulphate (used as fertilisers) and caprolactam (a raw material for synthetic fibres).[15]

From 1972 onward, the ANIC plant saw a series of accidents, which had the long-term effect of changing the collective local psychology and transforming residents into citizens of the 'risk society'. These accidents allowed Manfredonians to see and clearly perceive – by their noses, ears and hands – what was being produced within the factory besides salaries and income. Ammonia, arsenic, nitrous acid, sulphuric acid and other pollutants were visibly released in a series of fallouts amounting to several tons each, provoking collective intoxication, mass escapes and panic. The most serious fallout occurred on a Sunday morning in September 1976 (two months after the more famous Seveso accident), when an explosion in the arsenic column caused the dome-shaped roof to blow off the plant, falling on a shed on the opposite side. Soon after, people walking in the main commercial area could see a wide brown cloud coming from the plant and moving towards them, followed by a yellow slush that gently fell like snow all around. That snow was arsenious dioxide, and it was later calculated that some 32 tons of it had fallen on the town.[16]

The symptoms of widespread contamination soon became apparent: the day after the explosion, many barnyard animals died and large quantities of arsenic dust were found on leaves. In the following hours, the first hundred people were hospitalised with strong symptoms of arsenic intoxication. These were mostly workers from the plant and residents from the Monticchio neighbourhood, a former rural area surrounding the factory that had become a crowded settlement of 12,000 poorly housed people who had migrated from the countryside in search of jobs. The management of the ANIC denied the existence of any risk and put the employees back to work as if nothing had happened. The only action it took was sending in a special team of maintenance workers, to clean up, who were given no protection and had no idea what they were handling. These workers swept away the arsenic dust day and night so that the plant could resume regular operations the following Monday. Soon after the accident, in October 1976, six top managers from the ANIC plant were investigated for 'negligent slaughtering', but the preliminary inquest did not even get to the courtroom. In fact, the slaughtering was not evident, yet it would become so only a couple of decades later, when a number of workers who had entered the factory in the early 1970s came to suffer illness and death by a variety of serious ailments related to the acute arsenic intoxication of 1976.

The Manfredonia accident occurred when the golden age of labour environmentalism in Italy was coming to an end; moreover, it happened in the south. There was no SMAL in Manfredonia nor any public health officials, students or even unions willing to counteract ANIC's overwhelming contamination of both human bodies and social values. The politics of 'deceit and denial' were easily implemented in this case; in fact, the Manfredonia accident is still mostly unknown in the international literature and even to the Italian public. The existence of a well-consolidated knowledge/power assemblage connected to the Italian state was materially experienced by the victims of the 1976 accident in the form of delayed and misinterpreted data coming from labouratory tests and the deliberate manipulation of scientific standards with the aim of altering test results. Well-known and respected industrial hygienists, at the Labour Clinics of both Bari (the closest city) and Milano – all employed as consultants on ENI's payroll – denied public access to test results for nine precious days, then revealed levels of urine contamination from arsenic that were 20 to 50 times the maximum standard for several hundred cases. Local hospitals, however, were not able to receive so many people at once, and a number of victims were sent home having received no care. Company doctors decided to arbitrarily raise the levels of allowable urine contamination by 100 and 200 times in order to declare most of the employees 'able to work'.[17]

The 1976 accident gave rise to no wide or significant reaction from the community. The attempt to minimise hazards, by delaying test results and by recalling almost all of the workers to work, had the effect of reassuring a population still largely unaware of the real consequences of the contamination. Only a radical Left, grassroots organisation, Democrazia Proletaria (DP), attempted to keep public opinion alert to the 'lock on information' enacted by the government in Manfredonia – a practice already manifest in the Seveso experience. A few hundred people participated in a public demonstration set up by the DP some weeks after the explosion. The participants were not mere observers but people directly affected by the environmental consequences of the ANIC operation. These protests did not represent the voice of isolated and elitist environmentalists, but came from the world of work. Factory employees denounced having been sent to work with high levels of arsenic in their urine and no protection against the environmental contamination within the plant; local fishermen – a group that in the past had been strongly representative of the community identity and that continued to produce a significant part of the town's income despite the growing threat to their livelihoods from the ANIC plant – claimed that the Harbour Office had kept evidence of marine pollution in the bay area secret in order not to create alarm and disturbance in the local economy. Most interesting, however, are the voices that came from the Monticchio neighbourhood, where a Citizens' Committee for the Defence of Health was created and where a march of more than 10,000 on City hall began on 17 October. Nevertheless, DP was not a powerful organisation with thousands of affiliates, nor could its social influence grow much given its clearly declared loyalty to the extra-parliamentary Marxist Left.

Other political forces, including the Left parties, were largely absent from the social construction of community opposition to the plant, on this as well as later occasions. In understanding such a position, the 'job-versus-the-environment' discourse is probably the most relevant explanation: no party or union wanted in fact to be identified as contesting an agency offering employment, even if only 850 people effectively worked in the plant in 1976 and the total number in the following decades would never exceed a thousand.[18] Even more deafening – for what concerns us here – was the silence of the unions in Manfredonia. No SMAL was mobilised to assess the effects of the 1976 accident or the long-term effects of urea and caprolactam production on ANIC workers and their families in the surrounding area. Such a striking difference between the politics of the union confederation north and south is still in need of a historical explanation, hopefully achieved by future research.

Nevertheless, the lack of initiative on the part of (male-led) unions left open the possibility for another agency to come to the fore: women. It was a group of 40 local women, those most affected by the accident's fallout while living in proximity to the plant and the wives of its workers, who mobilised. Embracing an ecofeminist approach, they formed a Womens' Citizen Committee and succeeded in bringing the ENI group to court. Not the Italian court, though, but the European Court of Human Rights in Geneva, which, hearing the Manfredonia case in 1988, eventually came to rule against the company in February 1998. The Court declared Enichem guilty of moral damage by highlighting the relations between the toxic wastes and emissions from the plant and the women's private/family life. The ruling was centred on the 'right to know' – that is, the idea that the plaintiffs were entitled to access to information strictly concerning their own and their relatives' properties (house and body) and that the company had illegally withheld that information. The Court also declared the Italian state guilty for not protecting the plaintiffs from the violation of their privacy. The 'right to know' theory, however, does not imply the liability of a company (or the state) for the direct consequences of production. While the women of Manfredonia had asked for a huge, collective settlement for 'biological damage', the Court granted each plaintiff an individual sum as compensation for 'moral damage', for a total amount of one-fiftieth of the original request. Even more striking, the Court rejected the request of the plaintiffs that the Italian state be compelled to clean up the area, to establish an epidemiological study of the entire Manfredonian population, and to open an inquiry into the environmental impact of the Enichem plant.[19]

And yet something happened in Manfredonia that again set up the possibility for industrial hazard to translate into social, and legal, action. In 1995, a disabled and retired worker from the Enichem plant, Nicola Lovecchio, casually met a physician at the Labour Clinic of the University of Bari, Maurizio Portaluri, for a routine medical check. At the time, Lovecchio was already suffering the consequences of the strategy of denial played by company doctors from 1976 onward: he had lung cancer that, had it been diagnosed a couple of years before when it was already visible by accurate X-ray, could have been effectively treated. Lovecchio, instead, was declared 'able to work' until the cancer was widespread, and he died at age 49, 21 years after the accident of September 1976.[20]

Portaluri represented the 'democratic physician' the movement for workers' health in the 1970s had called for, being allied with the workforce against employers and company doctors. Some years before, he had read a dossier filed by the organisation Medicina Democratica about another Enichem plant located in Porto Marghera, near Venice. The dossier documented the investigation that another worker, Gabriele Bortolozzo, had started against management detailing the criminal responsibility of the company doctors for the death and disability of many workers from various forms of cancer, all related to the production of vinyl chloride (VCM) and polyvinyl chloride (PVC), as well as for environmental devastation in the Venetian lagoon.[21] Bortolozzo's investigation had opened up a new possibility for labour environmentalism in Italy, one that may be termed 'workers' epidemiology'. This led Portaluri to think that something similar could, and indeed should, be done in Manfredonia. Together, the physician and Lovecchio, before his illness overcame him, decided to carry out a bottom-up investigation: Portaluri asked for help from the 'militant' experts of Medicina Democratica (medical doctors, biologists, engineers), while Lovecchio interviewed his colleagues (and their widows), collecting memories of the 1976 Manfredonia accident and any relevant data concerning the work environment; he also solicited his fellow workers to ask the company for their clinical files. The final result of this research was a trial, involving hundreds of plaintiffs, a number of organisations, the town of Manfredonia and the Italian state, which, sadly enough, was concluded in March 2011 with the dismissal of all charges against the company.[22]

CONCLUSION

The stories told in this chapter offer particular insights into the historical agency of labour in environmental matters. In addition, it meets the call of scholars for an 'embodied environmental history'.[23] First, by combining with union action for the recognition of the objective value of workers' knowledge about industrial hazards, the militant new industrial hygiene of the 1970s translated into political change with general social, and environmental, impacts. Many improvements in occupational and environmental health, and even the encounter between Portaluri and Lovecchio at a public hospital in Apulia and all that came afterward, would not have been possible without that extraordinary season of Italian labour environmentalism, creating both the cultural and the material conditions for a public health system and for the enforcement of workers' right to know about the hazards of production. Second, and equally important, is the great historical significance of the workers' (bodily) knowledge of industrial hazards and the ways in which this could translate into political action beyond (and even possibly against) that of labour organisations. The case of Manfredonia, in particular, shows how the encounter between militant science and this embodied knowledge, and thus the possibility of socioenvironmental change, is not necessarily confined to the context of organised labour or even to the political hegemony of the unions, for it can arise from the initiative of individuals. Predicated by the Italian Left in the 1970s, and then abandoned under the pressure of economic recession, the Gramscian strategy of class solidarities became itself embodied in the story of a personal encounter: that between Lovecchio and Portaluri in the impoverished and heavily polluted Manfredonia of the mid-1990s.

One final point of methodological significance for an environmentally conscious working-class history emerges from the chapter. Environmental historian Arthur McEvoy once suggested that workers' bodies be seen as metatexts on which the political ecology of industrial societies has been written – and in many ways this perspective is reflected in the stories told in this chapter. But the chapter has also shown workers as self-reflective agents of environmental change and workers' bodies not simply as biological machines but as historical actors endowed with cultural and symbolic tools and producing not only commodities but knowledge and political agency as well. In the Italian case, the knowledge of work–nature relationships embodied by factory workers has become a powerful lever of environmental consciousness and action. In association with militant expertise, working-class people north and south, men and women, have historically acquired the ability to read the work environment and their own bodies, and take action for a stronger science and a more just society.[24]

3

Refusing 'Nuclear Housework': Sex, Race, Class, (Dis)ability, and the Political Ecology of Care

Over the past few decades, the environmental struggles of working-class, peasant, Black and Indigenous communities, and of women among them, have been investigated in political ecology and environmental justice research.[1] However, very rarely has this been understood as labour environmentalism, or else as an expression of people's relationship with the environment as mediated by their labour, and of the political potential of unwaged work in disrupting capitalist political ecology. Based on a historical case which illuminates the environmental significance of unwaged housework, this chapter offers a contribution in this direction, by highlighting a hidden dimension of the nexus between labour, gender and ecology in the Great Acceleration era.[2]

This chapter presents an original reading of the Wages for Housework (WFH) campaign as a type of labour movement, organising as mothers (and women) from low-income and marginalised social groups, and connecting their struggle against nuclear power to contemporary struggles around earthcare. I read this historical episode as an original and important contribution of the women's movement to environmental justice, which has remained largely unacknowledged so far. Most of all, however, this chapter offers new elements for reflecting on the political significance of mothering. Since, like other forms of labour, mothering embodies capitalism's ecological contradictions, its historical agency has involved countering the toxic, degrading effects of capitalist/industrial modernity upon working-class and racialised bodies, via both additional care work and struggle.

OPPOSING NUCLEAR POWER IN 1980s BRITAIN: THE HISTORICAL AND POLITICAL CONTEXT

On 21 December 1988, a group of women organised through the WFH campaign testified at a public Inquiry session held in Cannington (UK) against plans for building the Hinkley C nuclear power station. This testimony was structured as a collective objection from four organisations (the WFH campaign, Black Women for WFH, Winsivible – Women with Visible and Invisible Disabilities and Bristol Women's Peace Collective), each presenting distinct but correlated pieces of evidence, converging towards one claim: that the true cost of nuclear power could only be assessed by counting the unwaged caring work that is made necessary to counter its negative impact upon the physical and mental health of deprived communities.[3]

This was one of 182 sessions held before the Inquiry between October 1988 and December 1989, during which time an astounding number of 23,000 organisations and individuals registered[4] as 'objectors' to the application made by the UK's public agency Central Electricity Generating Board (CEGB) for building a third nuclear reactor at Hinkley Point, approximately 50 miles from Bristol city centre. Taking place less than two years after the Chernobyl accident, the Inquiry likely reflected a peak in public concern with nuclear power, as experienced by other European countries in the same period.[5] The aftermath of Chernobyl in the UK was in fact repeatedly mentioned in the WFH testimony as a precedent that demonstrated the incapacity of government agencies to deal with radiation risk, causing distrust and fear in the population. Nevertheless, this was not the central point in the WFH testimony, which focused instead on issues of 'differential vulnerability'[6] to radiation risk among deprived sectors of the population, and on the cost of this for women in terms of additional care workload. In other words, the WFH objection was motivated and argued in its own terms, based on the movement's unique perspective on nuclear risk as related to unwaged work, and structured by sexism, racism, ableism and poverty.

To be officially registered as a major objector, individuals and organisations had to submit a statement of case and proofs of evidence in writing before testifying. This required a tremendous effort in terms of time and resources invested in the process, which also weighed differently on social actors based on their respective capacities.[7] The fact that the WFH campaign decided to pursue this effort is thus significant, and testifies to the relevance that anti-nuclear mobilisation had for the movement's political vision, and their commitment to the anti-nuclear cause. In addition to this, the WFH campaign also mobilised to change the Inquiry process itself, by collecting a total of 13,000 signatures (two-thirds of which from women) on a petition for holding the Inquiry sessions in Bristol city centre (instead of in Somerset). The motivation for this demand, which was won in July 1989, was that of broadening the accessibility and thus inclusivity of the process for the people most affected by nuclear risk and damage, who – according to the testimony provided by WFH – were also the most socially disadvantaged: Black, immigrant, disabled people, and working-class communities of inner-city Bristol, as represented by their mothers.[8]

Overall, the Inquiry was a unique opportunity for the WFH movement to elevate the ecological dimension of their struggle, by demonstrating the unequal social impact of radiation risk and damage across lines of sexual, racial, ability and class discrimination. In doing so, they were influenced by and also actively contributing to the international ecofeminist movement, which was deeply connected with the long-standing women's peace and anti-nuclear arms movements, including prominent intellectuals and scientists.[9] One of these was the internationally renowned radiation and public health researcher Rosalie Bertell, a laureate of the Right Livelihood Award in 1986,[10] who had been studying the effects of low-level radiation on infants and children for a decade. She was, among other things, one of the contributors to a collection of ecofeminist and peace writings that also contained a contribution from WFH activist Wilmette Brown, who was chairing the WFH deposition in Cannington. Other authors included the renowned Kenyan scientist Wangari Maathai, leader of the Green Belt Movement (who would be awarded the Nobel Peace Prize in 2004); the US best-selling writer Marge Piercy, and a number of ecofeminist writers and activists from several countries.[11]

In short, the WFH objection to Hinkley C was part of a broader international women's movement for peace, ecology and public health. Its original contribution to that movement consisted in looking at peace, ecology and public health from the perspective of unwaged care labour. This reflected the WFH movement's core vision and struggle, which had been greatly reinforced since the final conference of the UN decade for Women in Nairobi had concluded with the passing of a recommendation – ratified by the UN General Assembly in November 1985 – to recognise and measure the economic value of women's unpaid work, especially that developed in the food and domestic sectors. The recommendation, contained in para. 120 of the final document of the Nairobi conference, had been won after ten years of WFH campaigning on the international level. At the national level, this had led to a motion on counting the contribution of women's unremunerated work to the UK's GNP, presented by the Labour Party in February 1988 and undersigned by over 100 MPs.[12] The times thus seemed mature for British society to accept the idea that unwaged carers counted, in both economic and political terms, and that their perspective should be considered when weighing the costs and benefits of public investments.

Interestingly enough, two trade unions figured among the most engaged objectors at the national level (in terms of resources and time spent in the process): the National Union of Mineworkers (NUM) and the Fire Brigades Union.[13] While NUM's opposition may be seen as motivated by sectorial interests (those of coal miners, in this case), a more attentive analysis may reveal a broader political scenario, heavily conditioned by the neoliberal project of doing away with union power, starting with that of the coal miners. Apparently, given the lower cost of coal in the same period, and the prospective raising costs of nuclear power due to the planned privatisation of the electric sector, Hinkley C could not be justified economically.[14] The main rationale behind its approval had to be political: officially, the government's decision for diversifying fuel sources for electricity generation, and specifically, for substantially reducing the quota of coal-fired power generation. Unofficially, a major advantage of the government's nuclear programme was considered that of 'removing a substantial portion of the electricity from the disruption by industrial action of coal miners and transport workers'.[15] In other words, Hinkley C appeared as 'the first power station to be built on purely ideological grounds, in that it was being used as a justification to destroy what was left of mining communities'.[16]

Since the Inquiry could not question the government's energy policy of reducing coal-based electricity generation, we can assume that the objection presented by NUM – based on the lower cost of coal-fired power – would be considered irrelevant. Nevertheless, the WFH deposition openly mentioned the claim that one main reason behind plans for Hinkley C was the government's intention to do away with organised coal workers, and that women in mining communities were mobilising against that decision: from their perspective, replacing coal with nuclear power made neither environmental nor economic sense. Alternative, more rational investments could have been made into 'clean coal' technologies, the WFH objection argued, thus sparing immense suffering to both coal-mining communities and to communities affected by radioactive pollution.[17]

ON NUCLEAR HOUSEWORK, AND ITS REFUSAL

Based on previous experience and on the available evidence, the objectors claimed to oppose Hinkley C because it would have reinforced women's unwaged workload, together with the deprivation of Black and disabled people and of working-class communities generally, especially that of inner-city Bristol; it would have increased inflation and cuts in social welfare; it would have augmented Britain's debt to the Third World,[18] in the form of uranium extraction and nuclear waste dumping; and for the inadequacy of compensation. All the above could be quantified, the objection claimed, by counting women's unpaid work.[19] In other words, counting 'nuclear housework' was proclaimed to be an effective tool for assessing the unsustainability of nuclear power production, by revealing a substantial part of its hidden costs and its unequal socioecological impact both locally and internationally. In the words of Nina López, testifying on behalf of the WFH campaign:

> Because it is primarily women who do the work of looking after everyone's health and welfare, for every radiation leak and for all the other physical and emotional damage resulting from Hinkley C, women would pay enormous and intolerable costs in increased unwaged work inside and outside the home – costs which are a yardstick for quantifying the human misery and ecological devastation of whole communities. It follows from this that the CEGB should and could have included women's unwaged work as a result of radiation pollution in their costing on Hinkley C.[20]

This claim was based on evidence which came, for example, from the World Health Organization, whose 1985 report had estimated that more healthcare was provided in the home by women than all healthcare services combined worldwide. Other relevant evidence was provided nationally by a 1980 DHSS report of the Working Group on Inequalities in Health, known as the Black Report (from the name of its lead rapporteur), and its updated version, known as the Health Divide report, published in 1988, and by a discussion paper on Deprivation and Ill-health, published in 1987 by the British Medical Association Board of Science and Education.[21] These reports highlighted the correlation between illness and social class, and specifically occupation and income. These structural health inequalities in British society – López argued – meant that the working class was particularly vulnerable to radiation risk, and that women in this class were bearing the cost in terms of additional workload. Due to pervasive and structural sexism, this workload fell mainly upon women, while its social value was unrecognised. Calculating this cost as part of the overall cost of Hinkley C was thus a unique opportunity to do justice to working-class women, and to uncover the true price of nuclear power production, making it prohibitive.

One important manifestation of sexism was the contempt with which the (then public) nuclear industry treated women, ridiculing their opposition as a manifestation of 'poor education'. This argument, made by the director of the Nuclear Electricity Information Group, a public authority of which CEGB was a member, responded to the results of an opinion poll launched by the Group in 1988, which showed that two-thirds of women opposed nuclear power (as compared to one-third of men) – after which the UK nuclear industry announced a £300,000 advertising campaign expressly targeting women.[22] The sexist diminishment of women's environmental activism as irrational was a long-standing strategy adopted by business, as famously shown by the worldwide famous story of Rachel Carson, the US biologist who had meticulously documented petrochemical contamination since the 1960s.[23] That this argument was still employed two decades later testifies to the persistence and power of sexism well beyond the emergence of feminism's second wave. However, it also testifies to the political potential of women's opposition to nuclear power, and to the danger women's peace and environmental movements were posing for the development of nuclear industry and armaments – as was also being shown by the Greenham Common's women's peace camp against the nuclear arms race.[24]

To counter the sexism of the nuclear industry, the WFH objection included the testimony of Suzie Fleming, an inner-city Bristol resident, member of the WFH movement and single mother of two children, one of whom was affected by meningitis. Fleming argued that women's concerns were based on firsthand experience with radiation through their work as caretakers in the most affected communities, such as those living nearby nuclear establishments or along the routes of nuclear waste transportation, and that these concerns had too often been proven to be true. It was the nuclear industry, she added, which had failed to convincingly disprove the possible correlations between low-level but continued radiation exposure and ill-health, especially the increased level of leukaemia and meningitis in the children of these communities. Citing the local press (*Daily Mirror* and *Bristol Evening Post*), she stated that in inner-city Bristol, which was crossed by the railway trucks linking Hinkley Point to Sellafield, where nuclear waste from the two existing power stations was stored, an incidence of leukaemia double the national average, and five times higher in children, had been reported. Radiation contamination had been found close to a primary school yard in Bridgewater, which CEGB had kept secret for five years before it was leaked to the press; additionally, all four leukaemia clusters in Woodspring were found to lie close to the railway line.[25]

In sum, women's concerns with nuclear risk appeared all too realistic. Based on her own lived experience, but also backed up by research and political activism on the international level, Fleming claimed that women routinely played the role of shock absorbers for nuclear risk in their families and communities. Even without considering the event of nuclear accidents, daily 'nuclear housework' consisted in making efforts to counteract possible radiation exposure by changes in the family diet; becoming alert to any sign of leukaemia or meningitis in children; and fighting to get the appropriate healthcare or compensation necessary to take care of ill family members. Additionally, radiation concerns implied the additional workload (and the related stress) associated with anti-nuclear activism. She described almost two years of organising by the Bristol Women's Peace Collective, culminating in a peace camp outside Avon's Public Protection Committee, to demand provisions for radiation monitoring and emergency procedures in the follow-up of the Chernobyl accident:

> This work included working our way through city council reports and the scientific reports by council officials; lobbying councillors ...; getting together to plan lobbying and demonstrations after a full day of housework, often with waged work on top; and then organizing childcare in order to meet; working to trust and understand each other as Black and white women from different communities; and working to circulate the information we had gathered to other women, including women who speak no English.[26]

It is important to clarify here that, although the WFH campaign talked of 'women' in general, identifying them with unwaged care work, the generalised use of this term was clearly a political figure, not different than the generalised use of the term 'worker', regardless of any social, economic or political differentiation, in the common language of the time. In the case of the WFH, however, this rhetorical generalisation was based on the awareness that women were the primary carers in every society. It was also counterbalanced in political praxis by the autonomous self-organisation of distinct groups of women, all of which belonged to low-income and marginalised communities. As explained by Fleming herself, 'it was in particular single mothers and women on welfare from the inner cities who led and persisted with this organizing' because, not being able to afford health protection, they demanded 'that local and national governments provided it'.[27]

The perspective of women with disabilities, testified by Claire Glasman from Winvisible – also an autonomous group within the WFH campaign – was then presented; Glasman highlighted the work of caring with people disabled by radiation, and the cost that the British government was saving thanks to unwaged carers, most of them being women. Based on the evidence on genetic malformations in newborns in post-Chernobyl Germany and in the newborns of UK nuclear test veterans, she stated that the cost of caring for people affected by the likely disabilities resulting from the new reactor at Hinkley C should be counted.[28] Further, she observed that people already disabled were disproportionately vulnerable to radiation due to their reduced mobility, depressed respiratory or digestive systems, compromised immune systems, circulation problems, proneness to infections. Radiation exposure could thus be directly correlated with increased self-care workloads for disabled people and their carers. Making things worst was the perverse prioritisation of expenses related to nuclear risk and radiation, with over £22 million having been spent in the past year to promote civil nuclear power, instead of countering the effects of radiation pollution and investing in accessibility of public spaces for people with disabilities. The very building in which the public Inquiry was held in Cannington – she observed – had no accessibility provision. This simple fact showed how disability was not even considered in public decision making, and how governments counted on the unwaged work of women and other carers to provide for basic services, so as to make cuts to public spending in welfare and rights. This had life-threatening consequences for many, and caused the rising unemployment of social service workers. All this, compared to the rising costs of producing nuclear power (as admitted by the same CEGB), and its prospective privatisation, made the government's choice to subsidise it even more politically unacceptable. Finally, she noted that evacuation costs in the case of accidents and fallouts must include the additional cost of evacuating people with disabilities, which, given the acute shortage of accessible vehicles and the need for experienced assistance, was likely to sensibly increase the overall cost of the new reactor.[29]

The WFH testimony was concluded by Wilmette Brown, a point of reference for Black Women for Wages for Housework. A former member of the Black Panther Party, Brown was a native of Newark, New Jersey (USA), an area that residents called Cancer Alley due to the high concentration of petrochemical plants.[30] After graduating at the University of California at Berkeley (class of 1968), she became a civil rights and anti-Vietnam war activist before joining the WFH campaign; she led the WFH delegation at the final UN Women conference in Nairobi, and was a joint coordinator of the King's Cross Women's Center in London, which hosted the UK branch of the WFH campaign and several other women's organisations.[31] She was the author of *Black Women and the Peace Movement* (1983) and of *Roots: Black Ghetto Ecology* (1986): based on her experience as a cancer survivor, this was a reflection on the nexus between capitalist development policies and the sickening of racialised bodies via degenerative diseases, what she called 'the malignant kinship of sex, race and class'.[32] At the testimony, Brown gave a detailed exposé of the extra radiation risk and damage suffered by Black and other people of colour, and by 'Black, ethnic and immigrant women' in particular. Racial discrimination was quantifiable, she stated, both in terms of resources of which people of colour were systematically deprived, and of the unwaged work spent 'in coping with and challenging the consequences of this deprivation'.[33] Research had shown how poor Black women were more vulnerable to ill-health than the rest of the population, and this must include increased vulnerability to radiation-induced cancer; further, the impact of racial discrimination on ill-health included increased exposure of Black people to nuclear pollution, nuclear worry and nuclear accidents. She also noted that racial inequalities had to be taken into account when calculating the cost of evacuating people from inner-city Bristol, where the police could not be counted on for evacuation purposes due to the distrust they had gained from the Black and immigrant population.[34]

Most of all, however, the problem racialised people faced in the UK was that experience had shown how their lives and health were 'even more expendable than most, literally worth even less, in terms of a hierarchy of monetary values, attached to employment, wages, housing and so on'.[35] This, Brown claimed, resulted from Britain's colonial and genocidal history, through which Black people had been established as a cheap or free source of wealth for the metropolis. Building Hinkley C would have simply added one more step onto 'Britain's lack of accountability to the international community on matters of pollution and otherwise',[36] what she referred to as 'Britain's debt to the Third World', a concept very proximate to what environmental justice activists and scholars have termed 'ecological debt'.[37] This consisted, for example, in dumping nuclear waste upon former colonised territories which could not refuse it because they were already impoverished and weakened by financial debt, thus turning Black bodies and territories into dumping grounds for the metropolis. It also consisted in forcing poor countries such as Namibia into provisioning uranium in unsustainable ways, with a heavy toll on the country's public and environmental health, as denounced by the Campaign against the Namibian Uranium Contracts in their 1986 report;[38] and supporting apartheid in South Africa, also in view of getting a share of its uranium resources. All this clearly pointed to how expendable Black lives were, abroad and at home, to the UK nuclear industry. Since a safe solution to the disposal of toxic waste was far from sight, she argued, the building of one more nuclear reactor in the UK was a matter of concern for people in the Third World, where radioactive waste tended to end up, and where already existing poverty and lack of primary healthcare made people more vulnerable to radiation exposure. This would add more workload on women, who were the primary carers of their communities, as abundantly demonstrated by 'the shattering burden' of surviving through the consequences of nuclear bomb testing in the Pacific Islands, mostly manifesting in miscarriages, birth defects and severe disabilities in the newborns.[39]

Finally, if counting women's additional workload, the inadequacy of compensation for radiation damage, as already experienced by the affected people, would be reinforced and increased. If nuclear industry workers had had to struggle for many years to win the right to claiming for compensation for themselves, Brown noted, imagine how hard it would have been for those 'unwaged women who, like it or not, are turned into nuclear workers in the community' to get their right to compensation recognised and paid. This was no trivial point, for an adequate compensation system, recognising the damage suffered by both 'productive' and 'reproductive' nuclear workers, would have made the cost of nuclear industry prohibitive, possibly 'pricing it out of the market'.[40]

'NUCLEAR HOUSEWORK' AS AN ENVIRONMENTAL JUSTICE ISSUE

Overall, it was their positioning at the margins of society, at the intersection of sex, race, class, (dis)ability and other discriminations, that gave the WFH movement its political authority and legitimacy, based on the claim of representing a perspective that was typically missing (or hidden) from public discourse, despite its correlation with wealth, health and wellbeing. Their initial statement clearly summarises this point:

> As Black, ethnic and immigrant women, women with disabilities, low-income mothers and single mothers, we bring our perspective and our protest against Hinkley C to this inquiry because all-pervasive sexism, racism, ethnocentrism, disability discrimination, and class discrimination, mean that we are among those least likely to be seen or heard, kept informed of danger or even considered for protection. In our experience even many who accept that economic and social exploitation are pervasive have trouble connecting this with nuclear pollution and nuclear 'accidents' ... Even without conscious targeting, those of us who are socially and economically disadvantaged are automatically targeted for the worst effects of radiation.[41]

Although the term 'environmental justice' never appears in their testimony, this was in fact highly consistent with the environmental justice approach as developed in Black working-class communities of the United States in those same years: denouncing the unequal burden of environmental costs upon Black communities, and demanding compensation for it, the movement against environmental racism exposed the entire system of waste production and disposal in that country, in the hope that this would have caused its collapse.[42] Similarly, in demanding total reparations for women's unwaged work, the WFH campaign aimed at causing the collapse of the nuclear industry, and of capitalist patriarchy more broadly. The similitude becomes clearer when considering a reference to the civil rights movement made by Wilmette Brown and repeated by Nina López during the cross-examination: Brown referred to Martin Luther King's claim that, despite the fact that no amount of gold could adequately compensate Black people for centuries of exploitation and humiliation, a price could be placed on the unpaid wages which were due to them.[43] For the WFH campaign, putting a price on unwaged work was not a purely symbolic gesture: compensation for unpaid care was (and still is) actively pursued by the movement in different forms, both as an immediate response to carers' material needs and poverty, and as a form of reparative justice. Similarly, the environmental justice movement has been using the instrument of legal action and judicial trials to make polluters accountable to low-income communities, and get the money which is due to them.[44] These movements know all too well that the money they might eventually get can never be enough, nevertheless demanding it is also understood as a way of increasing the power of workers, thus limiting the power of capitalism, cracking the system from the inside. And this is why their demand for money is not an individual affair, but a collective and politically charged one. This also applied to the WFH objection to Hinkley C. As stated by Brown in her testimony:

> We share Martin Luther King's view that even as we put a price on and demand reparations – the unpaid wages due to women for the nuclear holocausts which have already happened – our best protection is to break the nuclear chain, to oppose

> the building of any further nuclear power plants, demand the closure of those already built, oppose nuclear testing and nuclear waste dumping, and campaign for the abolition of the nuclear weapons of all States. We are fortunate to be part of an international movement engaged in doing precisely this.[45]

To conclude, the core argument of the movement's testimony can be summarised as follows: if quantified in terms of women's caring work as required for countering the effects of radiation leaks via diet and lifestyle changes, managing stress derived from radiation concerns, gathering information and mobilising about radiation risk, taking care of ill relatives and so on, the real cost of nuclear power was clearly unsustainable. In other words, estimating the cost of 'nuclear housework' was seen as an effective means towards opposing nuclear power production on the only terrain that really mattered in a capitalist economy: the monetary one. This argument was met with manifest contempt on the part of the cross-examiner, who tried to reduce it to an estimate of the *projected additional (marginal) cost* that would derive from the new plant, a cost which was easily discardable by arguing that the new reactor would simply replace the existing ones. The opponents had to repeatedly state that their argument was based on the *already existing* 'nuclear housework' imposed by the CEGB upon deprived communities of inner Bristol – and that this was what motivated women's opposition to the continuation of the nuclear industry. In short, the argument was that housework constituted a significant hidden cost of nuclear energy that, if properly compensated, and added to other hidden costs (e.g., that of properly disposing of nuclear waste), would have immediately put nuclear power out of the market.

At the same time, the claim was that the hidden cost of 'nuclear housework' only became clear by taking social inequalities into account, and by listening to Black and other working-class mothers. Their materialist demand – for 'nuclear housework' to be counted and compensated – was instrumental to building their political subjectivity: not as activists, on behalf of the non-human environment or of future generations, but as domestic workers refusing both existing and additional 'nuclear housework'. This was indeed a unique perspective and an original contribution given by the WFH movement to anti-nuclear and women's movements; most interestingly, however, this approach testifies to the movement's early perception of the political link between environmentalism and unwaged care work, a link which is highly relevant to the political ecology of the Great Acceleration era, and to climate justice activism today.

As the previous chapters have shown, the Great Acceleration was not a success story benefiting all humanity, but a highly unequal development process, which, in racialised and low-income communities, increased the amount of daily care required to cope with occupational hazards, pollution, loss of resources and environmental risk, ending up with an increased burden of work for the carers. This overburdening of socially necessary reproductive work – not as physiological, but as a social fact – has been key in linking women's activism with environmental activism, be it called ecofeminist or otherwise. This suggests that, throughout the Great Acceleration era, the environmental struggles of working-class, peasant, Black and Indigenous women have been a specific form of labour mobilisation, that is, as struggles of reproductive workers (domestic, care, and subsistence) over the rights and conditions of reproductive labour, starting with its adequate compensation, intended as a means towards increasing workers' control over the conditions for life on earth.

4

Taking Care of the Amazon: On Interspecies Commoning in the South of Pará, Brazil*

In May 2011, Zé Cláudio Ribeiro da Silva and Maria do Espirito Santo, nut collectors and members of the agroforestry project (Projeto Agro-Extractivista, PAE) of Praialta Piranheira in the Brazilian Amazon, were brutally murdered as a consequence of their engagement in protecting the forest from illegal logging and timber trafficking.[1] Making a living from a non-exploitative and regenerative relationship with the forest, and passionate about the defence of the rights of both Amazonia and its people, Maria and Zé Cláudio were among the number of earth defenders whose lives are being taken, year after year, for opposing the infinite expansion of global economic growth.[2] The struggles of peasants, fishers and Indigenous people in defence of the earth commons at the frontiers of resource extraction and waste dumping is now well documented and acknowledged.[3] Commonly understood as the 'environmentalism of the poor',[4] these struggles are waged by people whose lives and work are systematically devalued and discounted, but who have persisted in resisting capitalist industrial modernity and pursuing alternative forms of development based on earth-caring relationships. Their historical agency has significantly contributed to the political ecology of unwaged subsistence and reproductive labour in the age of climate change.

From a feminist historical materialist perspective, I see these subjects as part of the 'forces of reproduction', that is, unwaged

---

* This chapter offers a new and updated synthesis of distinct parts of Stefania Barca, *Forces of Reproduction. Notes for a Counterhegemonic Anthropocene* (Cambridge: Cambridge University Press, 2020).

workers whose labour in reproducing humanity and the conditions for life on earth is fundamental to capitalist accumulation, and thus embody a specific potential for subverting the system.[5] Capitalist/industrial modernity, I have argued, is built upon inequalities of class, race, sex/gender and species – in other words, master relations; these hegemonic relations are not natural, but have resulted from struggles, and are continuing to be contested and counteracted by other-than-master subjects worldwide. In short, capitalist/industrial modernity is not the only – let alone the best – possible way of organising social and ecological relations in the modern world.[6]

A historically significant terrain of anti-master struggle has been the defence and expansion of the commons.[7] As this chapter will show, the struggle for the commons includes both class and species dimensions: in its more advanced political forms, it fosters not only non-master relations among humans, but also non-master relations between humans and non-human nature. This, I will argue, is what we can learn from the historical experience of the struggles for the earth commons in the Brazilian Amazon, and from the lives and work of Maria and Zé Cláudio. As leaders of a community agroforestry project in the south of Pará, they were members of a broader historical movement, the *Aliança dos Povos da Floresta* (APF, Alliance of Forest Peoples), formed in the mid-1980s from a coalition of labour and Indigenous organisations in the Amazon region, which had achieved the institution of a new type of conservation units, the so-called 'extractive reserves' (*reservas extractivistas*, or Resex).[8] As mentioned in Chapter 1, the Resex constitute a path-breaking model of conservation, where subsistence-oriented labour, organised through commoning relations, is seen as the best guarantor against forest destruction. In short, the APF offers a unique historical example of the extraordinary anti-master potential embedded in the political unity of Indigenous and labour organisations.

This chapter will look at the story of Zé Cláudio and Maria as one that particularly illuminates the nexus between commoning and environmental care. In doing so, it contributes to research on the political significance of affective labour in community forestry, and in the struggle for the commons more broadly.[9] The chapter proceeds as follows: in the first section, I will locate the Praialta case within the historical context of the APF. This section will show how the APF's political project was centred not simply on the struggle against private property of the means of production and for territorial sovereignty, but also on principles of more-than-human, or interspecies commoning, inspired by Indigenous cultures in the Amazon region. The second section will then discuss interspecies relationships through the bodies and words of Maria and Zé Cláudio themselves, as rendered to us by the docufilm *Toxic: Amazon*,[10] describing their work as taking care of the Amazon in the community forestry project of Praialta. Building on feminist thought, the third section will contrast interspecies commoning and environmental care work with species supremacy as developed by capitalist/industrial modernity. It will argue that ecofeminist and ecosocialist thought matter to the Praialta story because they show how the Indigenous call for interspecies commoning, coming from different places worldwide, has resonated with other anti-master subjects from across the colonial divide. This, I believe, is a necessary premise for a global political alliance of anti-master subjects, the only hope we have for a future that is liveable for all.

The chapter concludes by sketching some of the most significant implications of the Praialta story for today's labour environmentalism.

THE ALLIANCE OF FOREST PEOPLES

Originating in the mid-1980s, the *Aliança dos Povos da Floresta* (APF) is a unique historical example of a political coalition between Indigenous and labour organisations, more specifically the *União das Nações Indígenas* (Union of Indigenous Nations of Brazil – UNI) and the *Conselho Nacional dos Seringueiros* (National Council of Rubber Tappers – CNS).[11] From the late 1980s, the latter turned itself into a social movement union representing all kinds of *extractivistas*, that is, subsistence producers and earth defenders from a variety of biomes throughout the country. This extraordinary political coalition resulted in the institution of a new kind of protected area, the 'extractive reserves', or Resex. Modelled on Indigenous reserves, and aimed at preserving both biological and cultural diversity, the Resex achieved recognition from some states of the Amazon region already in the 1990s; in 2000, they were recognised at the federal level, and incorporated in the National System of Conservation Units.[12] Throughout nearly five decades, 94 Resex have been instituted in Brazil, protecting both terrestrial and aquatic ecosystems via sustainable community management.[13]

By taking care of the forest, protecting it from the destructive impact of mining and industrial development, and granting the regeneration of soil, plants, water and biodiversity, the *extractivistas* – as Brazilian commoners call themselves – have significantly countered the degradation of earth systems caused by capitalist/industrial modernity. At the same time, and for the same reason, their work has been violently opposed, and their very existence threatened, by those forces of Brazilian society which see them as obstacles to accumulation and GDP growth. The assassination of Chico Mendes, founder of CNS – the national rubber-tappers union, now National Council of Extractivist Population – and founder of APF, in 1988 was only one among a long list of killings, systematically planned and executed along what Brazilian political ecologists have called the 'arc of deforestation' (*arco do desmatamento*); as a result, this is now rapidly advancing deep into the core of the Brazilian Amazon, driven by the increasing global demand for soy, meat, energy, timber, iron, aluminium and other commodities.[14] In short, the history of Resex is ridden with violence, and with repeated social pressures to reduce the number of reserves or convert them into 'Environmental Protection Areas', allowing the entrance of agribusiness.[15]

In the case of Indigenous people, this is only the latest chapter in a much longer history of colonial genocide, which was always related to the expansion of capitalist frontiers into their territories. As many Indigenous writers and activists have noted, one important component of colonialism was the annihilation of Indigenous ontologies and epistemologies, that is, the knowledge and belief systems on which their modes of re/production were based.[16] Resisting this cultural annihilation is today a fundamental dimension of Indigenous struggles for territoriality; at the same time, this is part of a broader struggle for a different conception of humanity, and for a different modernity, to emerge in response to the planetary challenges of our time.

This is what we learn from one of the historical leaders of the APF, the Indigenous writer and activist Ailton Krenak. Defined as 'one of the greatest political and intellectual figures [to have] emerged from the Brazilian Indigenous movement since the end of the 1970s',[17] Ailton Krenak was a member of Brazil's constituent assembly after the fall of dictatorship, and a close friend of Chico Mendes, with whom he helped to establish the APF.[18] In his latest book, *Ideias para adiar o fim do mundo* (How to Postpone the End of the World), Krenak criticises the modern concept of humanity as premised upon the idea that there is one right way of inhabiting the earth, carried out by an enlightened people, whose planetary hegemony is justified by its universal civilisational mission. He sees this hegemonic humanity, now represented by global institutions like the World Bank and the United Nations, as characterised by a false ecological consciousness, a pretence of being separated from and mastering its environment. Sustainability, Krenak argues, must be seen as a fraudulent concept produced by this fraudulent humanity; it was invented to justify their assault on Indigenous conceptions of nature. 'We accepted it', he writes, because 'for a long time, we got caught in this idea that we are humanity' and thus 'we got alienated from this organism to which we belong – Earth – and we turned to thinking that the Earth is one thing and we are another'.[19] The only ones who found it vital to keep themselves clinging onto the land, he writes, were those peoples (Caiçaras, Índios, Quilombolas, Aborigines) that remained half-forgotten at the margins of the world. Krenak calls them the sub-humanity – a 'gross, rustic organic layer' that is clearly distinct from the 'cool humanity' that predicates itself upon separation from the rest of the natural world.[20] He suggests that postponing the end of the world has everything to do with contesting the homogeneous vision of humanity, a leverage point from which to condemn and reject the kind of world that the UN wants to save, while building alliances with various peoples who are struggling for a world where bio/cultural diversity is respected and celebrated as foundational to the polis.

Mendes, on his part, was deeply influenced by the Indigenous cultures he had encountered while living and working as a rubber tapper in Acre. In fact, Krenak recalls him declaring that the *seringueiros* had learned 'our way of raising children from the Indians and from the forest itself', and that this had allowed them to 'attend to all our basic needs', creating a culture of their own, 'which brings us much closer to the Indigenous tradition than to the "civilized" tradition.'[21] For the *seringueiros*, who were migrant wage labourers mostly coming from the poorest and drought-stricken regions of the Brazilian North-East, encountering the Indigenous subject in the Amazon forest was an eye-opening experience. While initially pitted against their Indigenous co-workers by the rubber barons, who saw and feared the potential of a united workforce, the rubber tappers realised the enormous liberatory potential that resided in the Indigenous relationship with the forest. They understood how the agrarian reform as pursued by the Brazilian government was a false solution and an economic trap which would have forced them to either destroy the forest or to sell it. They saw that the Indigenous struggle for territoriality constituted a true and much better alternative – one which would allow them to get rid of their masters once and for all, by giving them access to the means of subsistence. In short, the rubber tappers had learned that another world was possible, and worth fighting for. This was not simply a contingent political strategy, however, but (as this chapter will argue) included a significant change of the self and collective identity of the *extractivistas*.

Summing up, the anti-capitalist capacity of the APF, its unique contribution to the history of labour environmentalism, rested upon a particular way of being within and relating to the earth, and specifically to non-human life in the Amazon forest. I propose to call this relation 'interspecies commoning', that is, a non-individualised system of accessing the land in which human and non-human nature are experienced as equally relevant components of social identity. I understand interspecies commoning as at once an onto-epistemology, a mode of re/production, and a political principle developed through the historical agency of APF. I see the principle of *florestania* (literally, forestzenship),[22] signifying how the forest and its peoples constitute a polis, a more-than-human community endowed with proper political subjectivity and equal entitlements, as strictly related to the practice of interspecies commoning.

Based on traditional Indigenous culture, the relational ontology of interspecies commoning has been politically represented by the CNS movement throughout the past four decades, and practised by thousands of commoners like Zé Cláudio and Maria. The next section will seek to tell this story from their own perspective.

THE PRAIALTA STORY

Maria and Zé Cláudio lived and worked as nut collectors in the Resex of Praialta Piranheira, in the state of Pará, which they had contributed to creating, and had both been elected leaders of the commoners' association at various times. They both identified as *caboclos*, a term probably deriving from Amazonian Tupi language that indicates a 'person having copper-coloured skin' or a person of mixed Indigenous Brazilian and European ancestry. Caboclos and Indigenous people form a large part of the 'peoples of the forest'. Despite having received limited formal education, Zé Cláudio and Maria had educated themselves politically in the tradition of the rubber tappers' movement. Once stablished in Praialta, Maria had gone back to school and pursued a master's in environmental education: she became an educator following the approach of Paulo Freire's pedagogy of the oppressed, and was active in the empowerment of rural women via agroecological projects. She and Zé Cláudio believed strongly in community agroforestry as a socially just and ecologically effective way of preserving the forest, and in the state's obligation to grant Resex communities the exclusive right to own and protect the forest from capitalist encroachment.

In Praialta Piranheira, the home of Zé Cláudio and Maria, people had come together to reclaim the possibility for themselves to re-/exist with the forest, as Maria used to say – that is, to reconfigure their livelihood and political existence as members of a forest community (*forestzenship*), rather than outside it. Their story is beautifully narrated by Felipe Milanez and Bernardo Loyola in the docufilm *Toxic: Amazonia* (2011), which helped persuade UNEP to award them the Hero of the Forest prize in 2012. The first scenes of the docufilm, shot in October 2010, show Zé Cláudio taking the reporters to see Majestade, the secular Brazil-nut tree (*castanheira*) that stood at the centre of the land plot that he and Maria had made their home. He could not possibly tell them his own story without also telling them about hers. We see him walking through a shady, lush vegetation, then stopping before a large tree trunk, his arm reaching out to it and his palm delicately touching the bark. 'This is Majestade, the pride of the forest', he announces, showing the width and height of the plant. 'If it were up to me', he adds, 'she would remain here for many years still.' He pauses, both his arms reaching out to the tree, both his palms touching the green stratum covering the bark: 'Even if she died, if something happened to her, this trunk will still be here', he says, before lowering his head and turning to look away.

When the PAE (agroforestry project) of Praialta Piranheira was created back in 1997, Zé Cláudio recounted, 85 per cent of the area was covered by native forest stands, mainly castanha (*Bertholletia excelsa*) and cupuaçu (*Theobroma grandiflorum*); little more than 20 per cent of it had survived by 2010, he claimed, parcelled in different places and surrounded by monocrop plantations. 'It's a disaster for those like me, the *extrativistas*', he commented. A Brazil-nut collector since the age of seven, Zé Cláudio self-identified with the *castanheira* and could not imagine his life without it: 'I live of the forest, and I protect her by all means', he claimed. Maria, who had also grown up with *castanheiras*, shared the same life project. She made it clear that it was a contested project, one that required political struggle and might imply losing one's life. 'Until there is a *castanheira* here, I am willing to fight. Until there is one, I'll give my blood for her', she declared.[23] She and Zé Cláudio saw no separation between their lives and that of their home forest: the *castanheira* was their 'companion species', and the *castanhal* (*castanheira* forest) the more-than-human, interspecies commons of which they were 'kindred members'.[24]

This project, however, required the expenditure of political engagement and active citizenship (forestzenship, more appropriately). With the advancement of capitalist/industrial modernity in the twentieth century, the *castanhal* had become a disputed forest, its re-/existence depending on social struggles opposing developmental plans based on the expansion of iron mining and other commodities. The region of Tocantins-Araguaia – where the PAE of Praialta was located – had a long history of violence against both the *castanhal* and its people – Gaviões, Aikewara, Xikrin, Parakana, Assurini, Kayapó, caboclos, peasants, nut collectors – a history of which Zé Cláudio and Maria were active subjects. Clearly, they feared that violence could happen to Praialta, to Majestade and to them. Maria's sister Laísa also lived in the PAE with her family and animals: her house, where she was interviewed by Felipe Milanez and Bernardo Loyola, was a simple shack surrounded by vegetation, with an annexed workshop. 'This is our home', she says, smiling, 'our paradise, where we live.' Laísa and her husband Rondon then take the two visitors to see the plot, showing them the plants that grow there – cocoa, cupuaçu, castanha – and they stop by a slim, smooth trunk. 'This is the Amazon's gold, mahogany', Laísa says proudly. After the assassination, they had to leave their home for a while; they were scared that something could happen to the family. But fear could not prevail, and they returned to their paradise after a few months. 'I would not trade it with anything, I really wouldn't', Laísa adds.

Castanheiras and nut collectors were members of a forest community born out of both hybrid labour and political struggle, whose permanence was threatened by the advancement of capitalist/industrial modernity. This is not a pristine wilderness to preserve, but a naturcultural terrain where the metabolic rift of the Anthropocene is contested and opposed by Indigenous and peasant populations that configure as forces of reproduction. Humans defend the *castanhal* because this is the nature they have materially appropriated through long-standing metabolic interactions, making it fit for their subsistence. It is this interspecies being, I argue, that attracts hatred and disavowal on the part of those subjects that identify with the capitalist project, which presupposes the 'death of nature' and its objectification into a passive, mechanical means to the production of value.[25] To Zé Cláudio, Maria and Laísa, the *castanhal* is a living entity whose existence is not easily distinguishable from their own: it is this obstinate refusal of objectification that make them seen as enemies of progress.

In this sense, Laísa's definition of the *castanhal* as *nosso paraíso* ('our paradise') evokes the contested but nonetheless real existence of an other-than-master world, in which people become free from exploitation and alienation together with rather than away from non-human nature. Liberation here assumes an interspecies meaning: there can be no true emancipation in a degraded and threatened environment where earth others are sacrificed. This, I believe, is an alternate mode of humanity that offers the greatest hopes for undoing the global planetary crisis of our times.

SPECIES SUPREMACY AND THE ALLIANCE OF ANTI-MASTER SUBJECTS

The killing of Zé Cláudio and Maria – along with those of many other *extractivistas* and Indigenous people practising interspecies commoning – must be placed in the long-term context of violent class and racial relations originating with European colonisation of the Amazon region, and continuing today with capitalist extractivism, the 'commodity consensus',[26] and the politics of GDP growth. Like all commoners historically, they have been seen as major obstacles to accumulation, and their elimination has been justified by the productivist ethos. Nevertheless, this picture would be incomplete without considering the violent species relations that are also built into the global master's system. As the previous section has shown, Maria and Zé Cláudio were not only struggling for egalitarian social relations, but also for egalitarian species relations. In other words, they were radically anti-master subjects.

It is important to recognise that the master's logic and system are also contested from within the Western world. A case in point is species supremacy[27] – by which I mean the ideology that legitimates violence against non-human life as instrumental to human progress: this ideology has been long contested by theorists and activists from the Global North. Socialist ecofeminists and materialist feminist authors, for example, have long argued that capitalist/industrial modernity cannot be understood without considering the specific type of relations that it has imposed onto the earth and non-human life since the early seventeenth century, with the concomitant development of private property, colonialism and heteropatriarchy.[28] In *Exhausting Modernity*, for example, Teresa Brennan argued that capitalism rests upon a 'foundational fantasy' of humans as self-contained individuals, inevitably denying the interdependency between human and non-human wellbeing.[29]

Clearly, this fantasy stands opposite to the principle of interspecies commoning – which it considers as its antithesis, the backward mode of humanity to overcome on the way to capitalist modernity. Val Plumwood argued for the importance of adopting a post-cartesian rationality that would enable us to recognise 'earth others as fellow agents and narrative subjects' within a 'dialogical conception of self' – a step towards enhancing 'interspecies communication'.[30]

Over the past decade, feminist scholars have greatly contributed to an exploration of transcorporeal affections, that is, the entanglements of human with non-human life within a common material reality that characterise the Anthropocene.[31] In her *Staying with the Trouble*, for example, Donna Haraway reminds us that 'No species, not even our own arrogant one pretending to be good individuals in so-called modern Western scripts, acts alone; assemblages of organic species and of abiotic actors make history, the evolutionary kind and the other kinds too.'[32] She proposes that, with its dramatic reconfigurations of the web of life, the Anthropocene should be seen as a time of passage towards something new, an epoch of 'multispecies ecojustice'. For this to happen, Haraway argues, it is necessary to invest our collective energies in a 'recomposition of kin' which might be allowed 'by the fact that all earthlings are kin in the deepest sense, and it is past time to practice better care of kinds assemblages (not species one at a time)'.[33] Most importantly, ecofeminism allows us to explore the intersected violence that oppresses living beings along lines of both social and species inequalities.[34] In being subject to different forms of oppression within a common matrix – colonial/capitalist/heteropatriarchy – the oppressed are seen as a more-than-human community of kindred beings.[35]

By interlacing with ecofeminism, historical materialism can help us make sense of interspecies-being as an insurgent practice of contesting the hyper-separations predicated by colonial/capitalist/heteropatriarchal modernity; that is, as an alternate mode of social emancipation and full realisation of human potential. While Marx intended species-being as a means to a dignified way of living through the affirmation of distinctive human potentialities – that is, the sensuous appropriation of non-human nature – interspecies-being could be understood as a recognition of the active role played by non-human nature in the realisation of human potential. As Gerda Roelvink explains, 'appropriation in species-being refers to the interdependence of the human species with "earth others" as they become part of, transform (and are transformed by), and thereby constitute humanity'.[36] Interspecies being, I suggest, could take full account of this co-constitution of humans with earth others as realised through a more-than-human labour process.

I would suggest that this perspective allows rethinking labour as an interspecies act, moving beyond masculine and human-exceptionalist notions of agency. Here I find it useful to employ Alyssa Battistoni's concept of hybrid labour, which she defines as a 'collective, distributed undertaking of humans and nonhumans acting to reproduce, regenerate, and renew a common world'.[37] As she puts it:

> Hybrid labor helps thread the needle between anthropocentrically instrumental and purely intrinsic value, recognizing the useful, material productivity of nonhuman nature without reducing it to the status of object or tool ... it aims to call a more-than-human political collective into being, and to propose a relationship to nonhuman nature grounded in interdependence and solidarity rather than unidirectional management, ownership, or stewardship.[38]

The concept of hybrid labour makes visible how the life project of Zé Cláudio and Maria was radically alternative to the valuing of non-human life as realised in old and new forms of capitalist/industrial modernity. It does so by allowing the conceptualisation of labour's potentialities for different ways of preserving and of valuing non-human nature. Praialta itself could be seen as a result of hybrid labour: as ethnobotanical research has shown, before Zé Cláudio and Maria went to live there, the *castanhal* had formed out of a process of interspecies becoming. Its geographical concentration in the south of Pará was related to the nomadic Indigenous habitations of this area, with their intensive use of the fruit as a source of protein. The *agroextractivist* life project consisted in both appropriating and reproducing this interspecies assemblage by making a living with(in) it.

CONCLUSIONS

Focusing on environmental care and interspecies commoning, this chapter has shed light on a distinct form of labour, whose history and politics are missing from the hegemonic Anthropocene narrative. This story offers a glimpse into the possibility, and possible outcomes, of an alliance among Indigenous and working-class people in search of alternate, non-capitalist, non-colonial and non-extractive ontologies and relations to earth. Rejecting simplified views of human identity as self-contained, and of human labour as inherently opposed to non-human nature, the chapter has shown how interspecies commoners like Maria and Zé Cláudio have developed significant potential for resisting species supremacy, and the degradation of earth systems that comes with it. This potential, I have argued, is key to subverting the master's system. In fact, species supremacy must be seen as a key feature of capitalist/industrial modernity, which could collapse without it. The chapter has also argued that interspecies commoning resonates with anti-master thinking and activism beyond the Global South, more specifically with ecofeminism. I would suggest that this resonance speaks to today's labour environmentalism in important ways, pointing towards the possibility of thinking of non-human nature as co-constitutive of the labour subject (rather than its 'other'), and organising workers beyond the wage relation, struggling for radically anti-master relations with the earth.

Illustrations

1a. Demonstrating for 'Jobs, health, income, environment', in Taranto (Italy), 2 August 2012. Photo by Domenico Grossi (*Liberi e Pensanti*)

1b. Marching towards an *empate* in Xapuri (Brazil), 1986. Photo from Marina Silva (licensed under the Creative Commons Attribution 2.0)

2a. *Comitato Cittadino Donne* (Women Citizens Committee) demonstrating for environmental and political clean-up, Manfredonia (Italy), late 1980s. Photo by Rosa Porcu

2b. A delegation from the *Comitato Cittadino Donne* of Manfredonia standing before the European Court for Human Rights, Strasbourg, 1988. Photo by Anna Guerra

3a and 3b. *Wages for Housework* campaign demonstrating against nuclear power in Cannington (UK), 1988–89. Courtesy of the Crossroads AV Collective, London

4a. Maria do Espirito Santo and Zé Cláudio Ribeiro da Silva at their home, in the Agroecology Project Praialta Piranheira (Brazil), October 2011. Photo by Felipe Milanez

4b. II National Meeting of Rubber Tappers and I Meeting of Forest Peoples, 1989. Photo from Instituto de Estudos Amazônicos (IEA)

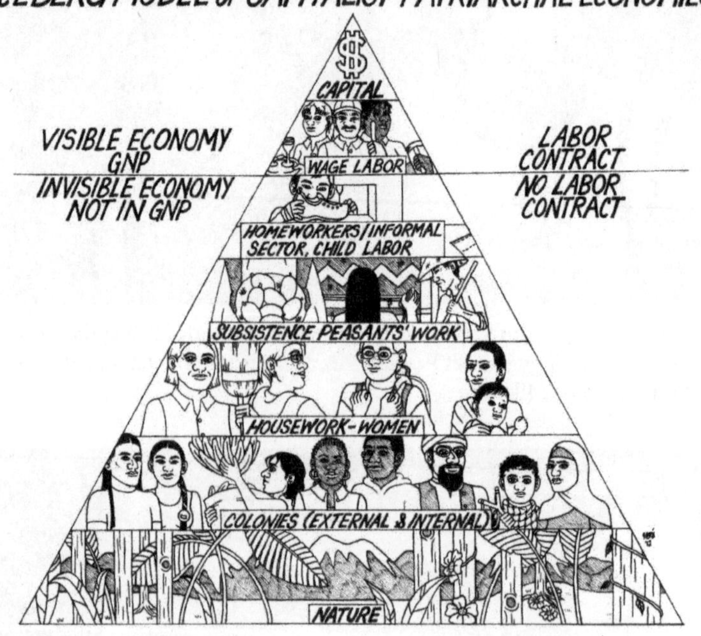

5. Illustration by Hannah Allen, based on Maria Mies' writings

6a. Demonstration of the Climate Jobs Campaign in Lisbon, 2018. Photo from Climáximo

6b. The Global Women's Strike marching at the XR 'big one' in London, April 2023. Photo from the Crossroads AV Collective

PART II

Political Ecology

5

Greening the Job: Trade Unions, Climate Change and the Political Ecology of Labour*

This chapter addresses questions at the intersection of green economy discourse (as well as practice) and work. The green economy, as developed in mainstream economic and political thought, is constituted by three agencies: nature, capital and labour. While scholars have paid due attention to the first two (i.e., climate change/resource exhaustion and capitalism), little consideration has been given so far to organised labour and to a broadly defined working class as those most affected by the process of greening the economy.

Broadly speaking, there is a strong need for political ecology to include work and labour more firmly on its research agenda. Trade unions and organised labour, different groups of workers and working-class communities rarely figure in political ecology narratives as key actors in environmental change and politics. Even environmental justice studies, be they focused on urban working-class neighbourhoods or on rural/Global South communities, pay insufficient attention to work-related aspects and to the relationship between occupational, environmental and public health.[1] The obliteration of 'class' and 'labour' in scholarship on environmental issues may reflect broader ideological divisions between labour and environmental movements, but this bias is much less significant in a field such as political ecology, traditionally conditioned by a

* This chapter is a slight revision of Stefania Barca, 'Greening the Job. Trade Unions, Climate Change and the Political Ecology of Labour', in *International Handbook of Political Ecology*, ed. Raymond Bryant (Cheltenham: Edward Elgar, 2015). Reprinted with permission from Edward Elgar Publishers.

leftist culture. Therefore, a (historical) political ecology of work could be undertaken in order to gain a better view on the political economy drivers of contemporary ecological crisis in its local to global implications.

Such an undertaking, as suggested in Chapter 1, could occur at three analytical levels: (1) the landscape, to encompass past human labour and social relationships incorporated into the land;[2] (2) the workplace, in its multiple forms and settings, as the context in which a working-class ecology unfolds, with its peculiar contradictions and political meanings; and (3) trade unions and labour organisations, as the place where a working-class ecological consciousness takes (or not) expression and evolves (or not) into political action.[3]

This chapter concentrates on the third level, and particularly on the mainstream labour movement's views on climate change and climate action. I first review an incipient scholarly conversation on labour's place in the green economy, suggesting that the ecofeminist economics perspective should be more fully integrated into it. Then, I provide a detailed discussion of 'climate jobs' and 'Just Transition' discourses on the part of international labour organisations, focusing on the One Million Climate Jobs campaign as originally developed in the UK and South Africa between 2009 and 2011.

CRITICISING THE GREEN ECONOMY: THE VIEW FROM LABOUR

Critiques of the 'green economy' are now well developed in political ecology,[4] while being contiguous to ecosocialist views expressed in such publications as *Capitalism Nature Socialism* or *Climate and Capitalism*. Differences exist, though, among that small group of writers who focus on the issue of 'green jobs', and specifically what is labelled the Green New Deal (GND) – that is, proposals for a recovery programme of public spending in the area of green energy and infrastructure, simultaneously aimed at creating jobs and saving the environment.

In the United States, for example, ecosocialist scholar–activist David Schwartzman, for example, argued that the GND should be embraced as 'a nexus of class struggle with the potential of opening up a path to ecosocialist transition on a world scale'.[5] In contrast, *Climate and Capitalism* author Don Fitz rejected both the GND and its global version – what UNEP calls the Global Green New Deal, launched in 2009 with financial backing from the OECD, WTO and IMF, and political endorsement by the green parties in the European Parliament. According to Fitz, the 'call for measures that will "stimulate job creation," in the absence of a parallel call to decrease the hours of work, is a *de facto* proposal for economic growth'[6] even as green production 'would mean a new influx of poisons from mining, processing, manufacturing, transportation and disposal within the GND's full life cycle'.[7] And those paying the heaviest price for that plan 'would be those it claims to help: low income communities of color'.[8]

A still different view was that offered by Jesse Goldstein, who distinguished between GND and the green economy: the former advocates a massive public works programme and expansion of the capitalist economy, while the latter encompasses more varied and imaginative proposals for private initiatives that may underpin a new economy.[9] Similarly, political ecologist Sara Nelson stressed the need for critical scholarship 'to reframe how we think about cooptation',[10] in the sense of looking for opportunities to experiment with anti-capitalist projects via the green capitalist movement (thereby reversing the usual direction of borrowing of ideas and practices).

This view connects well with the non-capitalist perspective presented in a 2014 special issue of the *Journal of Political Ecology*, and particularly in Boone Shear's article on the green economy. Due to its ability to 'accommodate elements of both Keynesian regulation and neoliberal development discourse', he wrote, the green economy may help facilitate a new historical bloc (in Gramscian terms) 'in which progressive interests could articulate around the idea that the construction of the green economy should benefit all social groups, as well as the environment'.[11] This project, however, could also entail 'a form of governmentality that seeks to discipline and produce people that will then reproduce capitalism'.[12] Working with union and community organisers of a Central Massachusetts Green Jobs Coalition (GJC), Shear (self-)reflected on what the green economy means to people at the grassroots level. Facing discrepancies between high hopes and often stark realities (e.g., fewer new jobs created than expected), he noted 'grim resignation' due to awareness that the green economy's ability to bring about a more just order of things 'is at the mercy of the [capitalist] economy, which is ultimately in charge'.[13] Capitalism is thus the equivalent of a 'Lacanian Big Other', the symbolic order that tells subjects how and what to desire, setting the conditions in which to play with such desires. Concurrently, this awareness dialectically generates a drive for transformative action beyond the limits set by 'green capitalism' in order to realise concrete possibilities for a non-capitalist order. A tripartite model resulted from the GJC debate: '1. Alternative economics – initiatives, enterprises, trade and finance that privilege community and ecological wellbeing over individual gain. 2. Resistance and reform – working against environmental degradation, social inequality, and poverty. 3. Social inclusion – efforts to end racism, sexism and other forms of oppression and exclusion.'[14] Inspired by Gibson-Graham's concept of 'diverse economy', Shear saw this discussion as a manifestation of how 'the green economy is a contingent, undetermined, economic space full of circulating desires, ideologies and fantasies, and a full range of capitalist and non-capitalist relationships and practices'.[15]

I return to this matter below, but wish first to enlarge our view in relation to work that has emerged, over the past decade, in different though converging areas of investigation. One of these is the sociology of labour, where scholars have maintained a critical if nonetheless open approach to green jobs strategies, highlighting both ambiguities and opportunities here. For example, Räthzel and Uzzell criticise such strategies for being largely directed at governments and businesses while involving 'unionists and workers as campaigners only, not as makers of their own futures'.[16] Instead, they invoke 'a trade union programme that makes use of workers' skills and knowledge to explore and design ways in which industries (and services for that matter) can be converted' as 'seeds of alternative forms of working and living in practice':[17] this, they claim, would be a strategy of 'revolutionary reformism' after Rosa Luxemburg. The authors also advocate a new scholarly approach called 'environmental labour studies', whose theoretical rationale is to understand how 'nature and labour are intrinsically linked and equally threatened by globalising capital', while 'the development of environmental trade union policies worldwide provides the empirical rationale'.[18] This argument was made in the introduction to a collection of essays on trade unionism and climate change, mainly focused on heavy industry – because this is the focus of most labour organisations in addressing the transition to a low-carbon economy (as shown below). While aware that this focus excludes 'the majority of workers in the countries of the North, who work in the service sector and offices, the requirements of which also make a significant contribution to climate change',[19] the authors still seemed to ignore the equally important exclusion of informal, unpaid and meta-industrial work done in agriculture, households and the care sector. Over the past decade, however, the ELS have evolved into a diversified field of research, which includes informal and reproductive work.[20]

According to socialist ecofeminist author Jacklyn Cock, a 'just transition to a low-carbon economy could contain the embryo of an alternative ecosocialist social order', provided that labour (and the Left) adopt it as a strategy 'for collective and democratic control of production, decentralised energy systems, production for social needs [rather] than profit, [and a] shift to agroecology'.[21] As she noted: 'the "green economy" is an empty signifier. Nowhere is it precisely defined. Everything depends on who claims it and gets to fill it with meaning'.[22] Indeed, such an economy is also place-specific, and may take on different forms depending on latitude or other geographical factors. And yet the global nature of climate change requires 'building transnational solidarity networks involving labour and environmental activists' that capitalise on traditions of 'social movement unionism' such as that developed in the anti-apartheid struggles of South Africa: these 'solidarity networks' are capable of mounting 'a powerful challenge to the neoliberal green economy'.[23]

The Ecofeminist Economics Perspective

The ecofeminist perspective on the green economy has been developed since the 1980s by diverse scholars in different fields interacting with ecosocialist and ecological economics scholarship. Feminist ecological economics, for instance, has reframed the economy as 'a complex of individual, family, community, and other interrelationships which each have economic and ecological significance', centred on the primacy of 'the work which takes place in households and communities'.[24] Notable early work here featured Maria Mies, Veronika Bennholdt-Thomsen and Vandana Shiva on the 'subsistence perspective', that is, a paradigm centred on that work that the universal spread of the wage-labour regime, and the associated conversion of all things and services into commodities, has undermined and made invisible in public discourse, including in the social sciences.[25]

For Ariel Salleh, 'any theorization of labor based exclusively on the experience of working men, is seriously deficient', as it does not account for the meta-industrial labour class, formed of those 'workers, nominally outside of capitalism, whose labor catalyzes metabolic transformations – be they peasants, gatherers, or parents', while 'supporting ecological integrity and the social metabolism'; she theorises this labour as 'relational, flow oriented, and regenerative of biotic chains'.[26] Meta-industrial work, which constitutes the largest labour class worldwide, coexists with capitalism, which could not do without it. By relying on knowledge from sustainability science and economics, with its heavy modernist bias focused on quantitative efficiency rather than eco-sufficiency, the labour movement remains 'locked into productivism and its technological fixes as much as the profiteering class is'.[27] This perspective is of fundamental importance to political ecologists wishing to make a theoretical contribution to discussions on the green economy. It invites scholars to extend 'sociological concepts of labour and value that evolved with industrial capital' by rethinking the relation between productive work and reproductive or regenerative (i.e., care) work, because 'a globally democratic resistance to capital calls for the recognition of "other" labors and the other value that they catalyze'.[28]

A crucial contribution here came from the feminist economic geographers Gibson-Graham and the Diverse Economies activist/research project. Going beyond the discursive violence not just of wage labour, but of the capitalist economy in general, they emphasised that 'non-market transactions and unpaid household work (both by definition non-capitalist) constitute 30–50% of economic activity in both rich and poor countries'.[29] They thus advocated 'a new economic ontology that could contribute to novel economic performances'.[30] This was based on the assumption that what is considered 'marginal' in mainstream economics and political economy – notably such things as care of others, consumer/worker cooperatives, community-supported agriculture, local and complementary currencies, the social economy/third sector, informal international financial networks, squatter, slum-dweller, landless and cohousing movements, the global ecovillage movement, fair trade, economic self-determination, the relocalisation movement, community-based resource management – are 'actually more prevalent, and account for more hours worked and/or more value produced, than the capitalist sector'.[31] This approach aimed to spur transformative research and politics in order to counter the tendency within the Left of marginalising projects of non-capitalist development to (at best) prefiguration as part of 'a politics of postponement'.[32] More recently, the authors have proposed an 'economic ethics for the Anthropocene' based on the need for a scholarship less focused on criticising techno-fix or value-shifting 'solutions' to the climate crisis, and more on reading 'the potentially positive futures barely visible in the present order of things, and to imagine how to strengthen and move them along'.[33]

Gibson-Graham's approach has become influential in political ecology, as evinced notably by a recent issue of the *Journal of Political Ecology* that invited engaged researchers to contribute to an ecological revolution by seeing current non-capitalist practices 'as part of a revolutionary politics'.[34] This approach is helpful in building a political ecology of labour that may help trade unions and the Left in general to envision alternative radical strategies beyond that of the 'green jobs' discourse under 'green capitalism'.

Critical scholarship on the green economy should thus take ecofeminist economics and political economy seriously, engaging in a fruitful conversation with its principles and political proposals. An ecological vision of work, from this perspective, must include both the blue-collar workers in heavy industry and the workers in agriculture, service jobs and meta-industrial labour. This vision would thereby crucially incorporate the large numbers of those who are excluded from the job market but nevertheless do most of the caring activities that make life on earth possible, allowing the reproduction of both human and non-human nature.[35] If failing to incorporate this perspective, the GND strategy implicitly reproduces the dominant gender hierarchy of the capitalist labour market by concentrating 'on the expansion of sectors such as energy and construction that are traditionally dominated by men', while 'unpaid female social reproduction work is thus silently accepted and assumed to be infinitely available'.[36] A detailed look at the mainstream labour movement's discourse on the green economy will make clear how this is indeed the case.

LABOUR ENVIRONMENTALISM IN THE AGE OF CLIMATE CHANGE

There are two historical phases in organised labour's environmentalism in industrialised countries: an earlier wave (starting with the birth of trade unionism) centred on the workplace and/or the living environment of working-class communities and the built environment at large, thereby stimulating convergence between occupational health and safety, on the one hand, and the protection of public and environmental health, on the other;[37] and a later wave (starting after the Rio 1992 Earth Summit) centred on concepts of 'sustainable development' and, more recently, 'green economy' and 'Just Transition' (JT). The latter concept expresses the idea that structural changes in the productive apparatus aimed at reducing its carbon content should not be paid by workers through job losses and destabilisation of local communities.[38]

Although the above picture does not account for differentiation among trade unions or geographical contexts, let alone between different periods in the industrial era, it nonetheless illuminates the long-term evolution of labour's perceptions of environmental issues, reflecting both societal and environmental changes occurring over the last 150 years. This evolution has encouraged coalition-building between labour and environmental movements, prompting formation of so-called blue–green alliances – that is, between unions representing blue-collar workers in heavy industry, transport and energy sectors, and environmental non-governmental organisations (NGOs). Examples can be found in the Blue Green Alliance (in the United States), which originated from collaboration between the Sierra Club and the United Steelworkers Union; or in the Global Climate Jobs Campaign, a coalition of grassroots campaigns involving unions and climate justice groups in different countries.[39]

Such alliances range from national to local level and represent the latest chapter in a long history of efforts designed to overcome divisions between red and green politics.[40] What is new about them is the growing political consensus that they embody regarding the need to tackle climate change, which has brought greening of the economy centre stage, while encouraging union mobilisation on environmental matters as part of a broader social agenda. This process has generated tremendous opportunities but also tensions in labour's environmentalism. Broadly speaking, within a common tendency of many groups to adopt a JT framework, important differences of interpretation persist: from a simple claim for jobs creation in the green economy to a radical critique of capitalism and a linked refusal to tolerate market solutions.[41]

There are clear political implications here. For example, in analysing US blue–green alliances, Gould and colleagues highlighted a loss of radicalism that they attributes to a post-9/11 politics centred on repression of dissent that has effectively marginalised anti-systemic strands in both labour and environmental movements.[42] At the same time, concepts of 'sustainability' and 'Just Transition' have become progressively more popular as common ground for a unified political strategy between the two movements based firmly on a non-anti-systemic platform promoting 'green growth' rather than on one based on more challenging environmental or social justice perspectives.

This evolution has reflected wider international changes. Especially important here has been the creation of the International Trade Unions Confederation (ITUC) as a result of the merger of the International Confederation of Free Trade Unions and the World Confederation of Labour in 2006, and the ITUC's launch of the first international labour's programme on climate change policies at the Trade Union Assembly in Nairobi in that same year.

This step stimulated creation of special union offices dedicated to formulating official positions on climate change, which have been increasingly geared towards the JT concept. It is thus worth briefly exploring this concept.

Setting the Agenda for Social Dialogue: The Just Transition Approach

Just Transition (JT) is defined by the ITUC as a 'tool the trade union movement shares with the international community, aimed at smoothing the shift towards a more sustainable society and providing hope for the capacity of a green economy to sustain decent jobs and livelihoods for all.'[43] According to former ITUC executive Anabella Rosenberg, it is 'a supporting mechanism of climate action, and not inaction. Just Transition is not in opposition to, but complements environmental policies. This confirms the idea that environmental and social policies are not contradictory but, on the contrary, can reinforce each other.'[44] Further, she claimed that the JT framework must incorporate: (1) sound investments in low-emission and labour-intensive technologies and sectors; (2) early assessment of social and employment impacts of climate change-related policies; (3) social dialogue and democratic consultation of social partners and stakeholders; (4) training and skills development; (5) social protection schemes, including active labour market policies, in order 'to avert or minimise job losses, to provide income support and to improve the employability of workers in sensitive sectors',[45] while addressing the consequences of climate change and extreme weather events on the poorest and most vulnerable; and (6) local analysis and economic diversification plans in order to help local governments to manage the transition to a low-carbon economy and enable green growth.

The JT strategy incorporates inputs from diverse stakeholders, including government (e.g., economic 'stimulus' policies), corporations (e.g., corporate social responsibility policies), academics as well as political leaders (notably propounding ecological modernisation discourses), and international organisations such as UNEP, OECD, EU, ITUC and ILO (e.g., their directives, reports and recommendations). All these stakeholder interests and inputs converged at the 2012 Rio+20 Summit in the form of a consensus on the need for greening the world economy in order to simultaneously save jobs and the climate, but without querying the current political economic system.

The UN's ILO (International Labour Organization) has been arguably most active in promoting this agenda. For one of its leaders, tackling climate change requires creating 'a global consensus that involves all stakeholders. Such a consensus will only arise if there is a seemingly "just" sharing of the burden in this battle to keep the planet hospitable to human beings.'[46] In the ILO vision, the role that trade unions need to play in this international process is to propose reduction targets for greenhouse gases obtainable through investment schemes that are driven by a preoccupation with jobs rather than a sole focus on reducing current production levels.

The ITUC has been a key partner of the ILO here. Such realpolitik reflected an awareness gained by the labour movement, especially after the 2009 Copenhagen negotiations, about the need to develop a 'job-friendly rationale' that would spur governments to act.[47] Overall, the ILO/ITUC vision of Just Transition emphasises a strategy of consultation and social dialogue, good governance and enhanced communication that serves thereby to hide lingering tensions and conflicts. It is mainly based on scientific claims coming from environmental economic and macroeconomic thinking; a notable role was played by a UK government-committed review known as the Stern Report (released in 2006), which demonstrated how the costs of fighting climate change through mitigation and adaptation investments were far less than the costs of inaction.[48] The ITUC then intervened with the idea that such investments were employment friendly. This focus persisted at high-profile international gatherings. For instance, in the lead-up to the UNFCCC conference in Bali (in December 2007), 'a strong statement was released by the trade union movement' raising concerns about the need for climate policies to 'become "job-literate"'.[49]

Clearly, the ITUC position reflects a vision of government economic intervention aimed at harmonising costs and distributing benefits of climate change policies among social parties. If carefully planned, the argument goes, both mitigation and adaptation policies can have positive social effects through job creation in 'infrastructure investments, such as the building of coastal defences, flood protection, drainage containment, road adaptation, etc.'.[50] Indeed, a new economic momentum is to be gained via investments that protect territories and populations from the adverse effects of climate events to come – the latter becoming, paradoxically, employment 'opportunities'.

This ILO/ITUC plan presented several problems. For instance, the impacts of massive infrastructure projects on local communities and ecosystems, however, were not considered; further, it did not take into account the fact that the Washington Consensus had forced virtually all governments to terminate social policies wherever possible and ignore (when not destroying) local economies, while adopting a competitiveness model based on ever lower labour costs and the hobbling of union power at the behest of global capital.

Another problem was that the potential to create jobs from investments in climate change mitigation (e.g., substitution of fossil with renewable energy) was (and still is) quite uncertain. Indeed, as such investment tend to be differentiated across time and space, it is all but certain to thereby create more (and not less) uneven development and inequality – something that is scarcely even considered in the plan. Instead, emphasis is placed on 'the expansion of renewable energies such as solar, wind, geothermal and agroenergy', with roughly 6 million jobs estimated to be created in solar power, 2 million in wind power, and 12 million in biofuels-related agriculture and industry by 2030 – a sector where every billion dollars invested is expected to create 30,000 new jobs such that it will be 'generating more jobs than the fossil fuel-based energy sector per unit of energy delivered'.[51] Nevertheless, the two empirical studies that were available on this aspect – 'one global assessment [prepared by UNEP, ILO, IOE and ITUC in 2008] and one [multicountry] macroeconomic study [i.e., concerning the European Union and produced in 2006]' – were both 'cautious regarding the net employment impacts of "green" policies'.[52]

The greatest problem with the ILO/ITUC plan is that it tends to gloss over how 'green energy' is notably about biofuels and hydropower production, whose sustainability and contribution to tackling climate change are (at best) disputable. Indeed, green labelling via the simple renaming of old activities (e.g., cash crop farming) as 'sustainable' might account for a very large share of the green economy balance sheet. Consider, for example, the 'green jobs' sector in Brazil – where biofuels (mostly produced from sugarcane) represent more than 50 per cent of total employment.[53] Further, as a dedicated literature has amply demonstrated, labour conditions in the *canaviais* (sugarcane plantations) are way below international standards, even as basic human rights are violated; meanwhile, mechanisation has meant that thousands of workers over the past decade have become unemployed without compensation or alternative employment;[54] in addition, sugarcane monoculture and processing constitute serious environmental and public health threats while replacing local food production, prompting in turn environmental conflict with local communities and Indigenous people.[55] And yet the Brazilian biofuel industry garners ongoing government, trade unions and ILO support due to its 'green energy' producer status – something that is today only reinforced further in order to sustain its competitiveness in the face of the recent discovery of huge offshore crude oil deposits.[56]

A big part of the problem in all of this is that, in assessing policies and envisioning solutions, both the ITUC and ILO clearly prioritise academic research over politically minded work. Thus, social movements and political parties, ecological distribution conflicts, social metabolism or radical perspectives such as degrowth, ecofeminism or ecosocialism had been largely left out of this conversation. While research questions were formulated on the base of 'knowledge gaps' in terms of foreseeing impacts of climate policies in different national contexts, economic sectors, and so on, none of the ILO/ITUC statements implied that interlinked ecological/economic crises arise from the current global politico-economic system, let alone that addressing the former necessitates transforming the latter. On the contrary, that system can apparently reform itself from within by way of a highly unlikely international coordination of different national schemes, each calibrated to that nation's development stage. Such coordination is unrealistic, as the failure of carbon schemes proves.

Indeed, in placing unconditional faith in 'green growth', the ITUC and ILO seem unaware that such growth is already taking place, not despite the economic crisis but precisely because of it: it is how capital produces new possibilities for accumulation, notably via investment in new technologies (e.g., renewable energy) and markets (e.g., carbon), just as the Second World War and postwar reconstruction were capital's way out of the Great Crisis of the 1930s. Parallels could be drawn between the staggering number of deaths and sheer amount of destruction that were required in that war with those that are already happening due to the combination of extreme climate events and long-term changes in resource availability and ecosystem functioning.

Moreover, nothing links the (purported) greening of the capitalist economy to decent work conditions and stable employment for labour. On the contrary, employers may well take advantage of this transition or 'restructuring' to eliminate residual workers' rights, as the case of the Brazilian biofuels industry makes clear. True, perception of such risks is not completely absent from supporters of the ILO/ITUC position. Thus, for example, the possibility exists 'that the winning tender may well be the one which pays the lowest wages, does not offer safety equipment or coverage for accidents, and which has the largest proportion of informal workers, for whom no tax or social security is paid, and who are not covered by any legal or social protection', a possibility increased by the tendency of public authorities towards 'outsourcing public and support services via contract, and financial investment in public–private partnerships'.[57] Yet the response tends to be unconditional faith in international conventions such as the ILO No. 94, which sets out 'a universal labour standard in the area of public contracting'.[58] Such faith, though, flies in the face of evidence that such conventions are notoriously ineffective and indeed blatantly ignored by most of the private sector.

Reclaiming the 'Green Economy'? The OMCJ Campaign

The most interesting example of JT as a labour-based coalition strategy is given by the One Million Climate Jobs (OMCJ) campaign. Launched by a UK coalition of trade unions oriented towards 'green growth', before also being claimed by a South African labour/environmental/social movement coalition in 2011, it originally promoted a Keynesian investment scheme designed to create 'climate jobs'. The latter are distinct from generic 'green jobs' insofar as they specifically seek to produce drastic cuts in the emission of carbon dioxide, methane and other greenhouse gases. In the UK case, these proposed climate jobs were envisioned in the greening of the electricity, construction and transport sectors – the most obvious initial target of climate action. Aimed at creating blue-collar jobs in these sectors, the campaign thereby remained silent on agriculture – a striking omission given that agriculture-related deforestation and grazing produce vast quantities of methane gas that in turn account for about half of total greenhouse gas emissions at the global level. While most such emissions are not produced in the UK (where, it is true, agricultural employment is today much reduced from the past), there should nonetheless still be an accounting of the UK's contribution towards the carbon footprint and social impact of the global agriculture economy, given its substantial agricultural imports. Likewise, the campaign did not pay attention to care, reproductive and domestic work, which are vital to overall social and economic wellbeing in any economy (i.e., green or otherwise).

The conspicuous lack of any anti-systemic language within the UK OMCJ campaign was not surprising, given that the entire JT discourse on which it was based aimed at consensus-building across trade unions, environmental NGOs (ENGOs) and governmental agencies at the national and international level. And yet, as discussion of the South African version of the OMCJ campaign next shows, the campaign has indeed the potential to mobilise different, more critical and ultimately transformative approaches.

According to Jacklyn Cock, South Africa embodies two global crises – rising socioeconomic inequality and climate change – thus experiencing tremendous tensions between official commitments to decarbonise the economy and to reduce poverty (including energy poverty): this has pushed the labour movement to uniquely orient Just Transition towards 'demands for deep, transformative change meaning dramatically different forms of production and consumption'.[59] Such change requires 'an integrated approach to climate change, unemployment and inequality, as well as a rejection of market mechanisms to solve these problems. Unlike some other formulations of the green economy, in this model the link between social justice and climate change is acknowledged, and the need for radical, structural change is emphasized.'[60] This approach to JT is driven by an anti-capitalist stance that worries that a decarbonised economy might simply reproduce current relations of power and inequality, based as it is on highly 'conservative' thinking surrounding sustainable growth and 'financialisation' of ecosystem services.

South African unionists developed this critical perception in light of their recent experience in the sector: specifically, after signing a Green Economy Accord in November 2011 based on a 'social dialogue' approach that bound government, business and labour together in the planned creation of thousands of jobs in a new green industrial base. Noting the many flaws and limitations in practice with this Accord – inflated claims not supported by evidence, persistently low standards and wages, job losses, and so on – they realised that these deficiencies derived from the fact that 'green jobs (e.g., in the privatised renewable energy programme) were driven more by the interests of the market rather than by social needs.'[61] Consequently, the national unions federation COSATU (representing 20 South African trade unions) resolved to adopt a Climate Change Policy Framework of 15 principles, among which were identification of capitalism as the underlying cause of global warming and (hence) rejection of market mechanisms to reduce carbon emissions (i.e., carbon trading). Departing from the ILO/ITUC definition of JT, COSATU argued that this concept must be 'developed further to fully incorporate our commitment to a fundamentally transformed society.'[62] Similarly, the National Union of Metalworkers of South Africa (NUMSA), one of the biggest COSATU affiliates, rejected green jobs as a component of a new green capitalism in favour of an alternative vision of JT, 'based in worker controlled, democratic social ownership of key means of production and means of subsistence.'[63] The Food and Allied Workers Union (FAWU) also expressed support for 'class understanding of a just transition to a green economy' and for 'radical alternatives to industrial agriculture, particularly agroecology.'[64]

Concurrently, COSATU and the National Council of Trade Unions (NACTU) joined with ENGOs and diverse social movements to launch a South African version of the OMCJ campaign in 2011. Aiming 'to exclude attempts by capital to use the climate crisis as an opportunity for accumulation', the campaign was strongly influenced by environmental and climate justice organisations, while also being based on 'a number of prefigurative projects in order to demonstrate the viability of policy proposals' put forward by the campaign.[65]

In the South African vision, the shift to renewable energy forms part of a wider transition towards publicly owned small-scale and localised energy production and autonomy at the household/village level. Firmly under decentralised community participation and control, this arrangement would grant everyone access to energy through low prices.[66] A similar approach is applied to food production and distribution, where a shift away from industrialised agriculture towards agroecology is envisioned, since climate change is understood as a seriously aggravating factor in food insecurity. As Cock concluded: 'While there is no consensus, there are elements of the transformative model of the green economy within the labour and environmental movements' and 'the embryonic alliance that is developing between these movements is a hopeful site for an effective challenge to the "corporate capture" of the green economy discourse'.[67] This in turn may generate transnational solidarities between labour and environmental movements.[68]

According to Italian scholar Emanuele Leonardi, though, the 'explicit and welcome politicization' of the South African OMCJ campaign with its 'complete rejection of any technocratic rhetoric' was nonetheless impaired by its heavy reliance on the state as the only social agency opposed to capital and the market.[69] He observed how the campaign assumed that local economies and communities must be protected from the potentially adverse effects of international trade rules as well as transnational corporations by state policies such as 'subsidies to local producers, non-price-competitive contracts, and import tariffs to help make foreign products uncompetitive'.[70] The problem, according to this author, 'does not concern the necessity to limit the market's all-pervasiveness but, rather, the very possibility that such a crucial task might be performed by the contemporary, heavily neoliberalised state deeply entangled in the cogency of the carbon trading dogma'. Instead, 'the prefigurative dimension of the *OMCJ* campaign would benefit from a non-state-based perspective such as that grounded on the notion of the *common/s*'.[71]

While giving such criticism its due, it is important to note positive features in the South African OMCJ campaign. For example, it was the only initiative, among those mentioned herein, of a large-scale green jobs coalition that included 'community caregivers' as its most relevant employment sector, foreseeing up to 1.3 million jobs to be created in domestic/healthcare, land restoration and urban farming; similarly uniquely, the programme was written with a significant contribution from women (i.e., 16 out of the 36 authors).[72]

CONCLUSION

On 21 September 2014, a People's Climate March of 400,000 participants took place in New York City – a demonstration of mass discontent with climate negotiations and injustices. Successful in getting such a sizeable population on the streets despite contradictions and criticisms surrounding the campaigning process itself, the march marked a historic convergence of disparate social movements and interests, including labour movements, towards the common goal of reducing CO_2 emissions in order to save life on earth. As with other historical cases (e.g., the nuclear disarmament campaign), large coalitions were enabled by a common survivalist perspective, prioritising mass mobilisation as the only way to overcome political and economic divisions at the decision-making level. 'To change everything we need everyone', the march slogan claimed.

Extensive involvement of different trade unions here reflected such a 'we need everyone' perspective, with the latter accommodating everything from radical anti-capitalist to green capitalist visions. This episode epitomises the sorts of challenges and opportunities that arise from the current climate/economic crisis, and that must be faced by organised labour, as this chapter has suggested. This topic also prompts intriguing research questions for political ecologists. Organised labour is indeed a vast and varied reality worldwide, whose diversity is scarcely covered in the present chapter, where the focus is on the official positions of large mainstream organisations such as the ITUC.

At the same time, there is a wide array of campaigning and organisational politics, including that of a more radical persuasion that was only hinted at here. Thus, discussion of South Africa's OMCJ campaign gives us an initial sense of the possibilities for reclaiming Just Transition and even Green New Deal strategies from a radical political perspective, filling them in with new anti-capitalist and (possibly) ecofeminist meanings. At the same time, one way of investigating organised labour's role in climate change politics is to look at the diversity of labour/environmental strategies occurring around the world today, linking them to different economic and place-based geographies as well as to shifting political opportunity structures in their continuous struggles to renegotiate social and ecological practices at the local and/or national levels. A feminist perspective on the possibilities of becoming – as offered by Gibson-Graham and the Diverse Economies research collective, for example – is a great point of departure for political ecologists who aspire to contribute to an 'ecological revolution' by researching, helping to develop, and intellectually supporting alternative visions and practices of work–nature relationships, of which organised labour can and ought to form part.

6

Labour and the Ecological Crisis: The Ecomodernist Dilemma in Western Marxism(s)*

INTRODUCTION

This chapter offers a critique of what I consider a major trend in the environmental politics of the neoliberal era in Western Europe: the political convergence between labour and mainstream Ecological Modernisation, what I will call labour's ecomodernism, and its contemporary divergence from anti-capitalist ecological movements. This pattern has become dominant in a historical context marked by a generalised decline in labour's representativeness and political power, both at the trade unions level and at the level of an almost generalised electoral defeat of the radical Left,[1] as well as by a wide adoption of neoliberal policies in Western European countries. At the same time, labour's ecomodernism hides important internal fractures and ecological contradictions: on the one hand, in the wake of increasing unemployment levels, a number of different sectoral unions and political parties on the Left continue to support fossil fuels and the opening of new extractive frontiers (from gold mining to fracking to coal itself); on the other hand, labour's endorsement of ecomodernism has been confronted by grassroots resistance against new 'clean energy' projects, such as, for example, wind farms and large-scale solar power plants, energy-from-waste facilities and high-speed railways. These divisions complicate

---

* This chapter is a slight revision of Stefania Barca, 'Labour and the Ecological Crisis: The Eco-modernist Dilemma in Western Marxism(s) (1970s–2000s)', *Geoforum* 98 (January 2019): 226–35. Reprinted with permission from Elsevier.

immensely the effort to delineate a red-green agenda, or even to understand where the front is located in the current ecological class conflict.

Addressing this conundrum, I argue, requires us to develop (a critique of) the political ecology of labour, that is, a material historical analysis of the internal relations between labour and ecology, focusing on the ecological significance of work and the political implications of workers' interaction with nature, both theoretically and in the historical praxis.[2] My critique situates itself within the Marxist ecofeminist perspective on labour and working-class agency. Adding to the notion of 'metabolic rift' as proposed by J.B. Foster,[3] Marxist ecofeminism focuses on 'the forces of reproduction', emphasising the 'metabolic value'[4] produced by forms of labour that lay beyond conventional (Western) understandings of the term. According to Ariel Salleh, 'metabolic value' indicates 'a subliminal "other" sphere of labour and value', that produced by the 'peasants, mothers, fishers and gatherers working with natural thermodynamic processes who meet everyday needs for the majority of people on earth'.[5] Necessary to industrial production and exchange value, these workers – she writes – typically inhabit 'the margins of capitalism–domestic and geographic peripheries', and thus they are 'unspoken, as if "nowhere" in the world-system'.[6] Naming them 'meta-industrial labour', while noting that they form the majority of the world's working class, is for Salleh an important way to make their work visible, and to value their contribution as 'rift-healing', that is, contrasting the degradation of bodies and ecosystems put in motion by industrial production.

Salleh's concepts of meta-industrial labour and metabolic value build upon a materialist ecofeminist approach,[7] which aims at a reconceptualisation of political ecology via a socialist/feminist dialogue.[8] I find this approach extremely relevant to a reconceptualisation of labour environmentalism in both its historical and present forms. Seeing the labour/ecology nexus from the vantage point of Marxist ecofeminism, the crux of the matter for a critique of labour's political ecology becomes: what forms of work and what working subjects are included in labour's ecomodernism? And, more in general, what counts as labour in labour environmentalism? This perspective would allow us to broaden the scope of labour environmentalism by developing a decolonisation of labour, both as concept and praxis, thus rendering visible its potentialities as an agent of ecological revolution.

Building on this approach, this chapter will develop a critique of labour's political ecology in Western Europe in the last quarter of the twentieth century. My intent is to reflect on the historical dialectical process by which Western Marxism confronted the ecological crisis, focusing on how it conceptualised labour and working-class agency in respect to ecology. My argument is that, though becoming increasingly aware of the constraints that prevented industrial labour from developing a proper response to the ecological crisis, European labour movements were incapable of developing a clear vision of the rift-healing agency of reproduction work (or meta-industrial labour) and of the need for uniting these two forms of labour subjectivity in a global solidarity alliance.

My narrative is built around the work of four public intellectuals – Laura Conti, André Gorz, Raymond Williams and Maria Mies – whose work represents different strands of Western Marxism, and whose influence extended across various social movements not only in the respective countries, but also internationally. Reading these authors against the respective historico-political background, and confronting them with one another, I believe we can: (1) discern the conceptual trajectory through which the labour movement of Western Europe came to embrace ecomodernism; (2) understand how labour's ecomodernism is distinct from the capitalist version of Ecological Modernisation; (3) develop a constructive critique of labour's ecomodernism from a materialist ecofeminist perspective. The next section will briefly delineate the current positioning of labour organisations within the contemporary political ecology scenario, and then introduce the research hypothesis and narrative that will be developed in the rest of the chapter.

LABOUR IN POLITICAL ECOLOGY

The first step for a critique of the political ecology of labour is that of defining labour and environmentalism as two composite fields of political action that are highly differentiated within themselves, and whose areas of intervention are overlapping in several ways.

This approach in turn builds upon an understanding of nature (and the ecological crisis) not as a self-evident thing, but as a contested concept that crosses through labour and environmental fields, being conceptualised and acted upon in different ways within each.

That environmentalism has never been a homogeneous movement, and that different souls have coexisted within it is a key understanding in contemporary Political Ecology.[9] Two forms of environmentalism, Ecological Modernisation (EM) and Environmental Justice (EJ), matter particularly in discussing labour environmentalism. The first, which now dominates environmental discourses in mainstream ENGOs and in global environmental politics (including climate negotiations), was originated as a North European stream of social theory in the early 1990s.[10] It offered an optimistic, win-win vision of environmental reformism as an effect of techno-fixes coupled with market incentives, which would come about as the result of a shift to post-materialist value systems in industrial economies. Like in most contemporary research on environmental politics, the 'post-materialist' concept was associated with a post-class, even post-political understanding of environmentalism. A highly contested theory, EM has nonetheless conquered centre stage in environmental policy-making at various levels thanks to its perfect fit with neoliberal environmentality.[11] As Maria Kaika has convincingly shown, EM has reached global dominance by being incorporated in the UN Sustainable Development Goals, where it has locked the debate on sustainability within 'the false dichotomy of market efficiency vs public accountability'.[12] Even though its age of innocence is now over, after its complicity with 'new forms of displacement and "environmental/ecological gentrification"' has been amply documented, EM now dominates the New Urban Agenda of the UN. This happens, according to Kaika, because the decision makers have chosen to ignore the voices of those urban communities and movements that are expressing dissenting, alternative visions of sustainability, geared on the praxis of 'commoning'.[13] A similar argument has been made by Goodman and Salleh regarding UNEP's official position on climate change.[14]

On the opposite front, global EJ[15] represents a subaltern and grassroots perspective which is gaining momentum in climate mobilisations, growing more self-conscious and poignant in identifying capitalism as the real culprit of the current ecological crisis – as in Naomi Klein's *This Changes Everything*, and in the climate justice movement more in general. The non-compatibility between this perspective and that of EM emerged with striking clarity at the Rio+20 Earth summit of 2012, where the official final declaration: 'The future we want' – a univocal endorsement of EM via a 'green growth' agenda – stood opposite to the alternative document approved by the Peoples' Summit, 'Another future is possible'. Seated at the official table and undersigning the UNCED declaration, the vast majority of labour organisations positioned themselves firmly within the first camp.

How should we make sense of this political positioning of labour with ecomodernism? According to Goodman and Salleh, the Rio Summit signalled the existence of a global counter-hegemonic bloc, formed of 'meta-industrial workers – urban women carers, rural subsistence dwellers, and indigenes' who represent the categories of workers hardest hit by the metabolic rift. These authors see the Peoples' Summit as an important step in the process of this global 'class' becoming self-aware of its political and ecological subjectivity. While largely agreeing with this vision, I believe we should not forget that the division between the EM and EJ blocs is internal to a broadly defined sustainability front, which stands opposite to the continuation of fossil-driven economic growth (with the recent addition of new extractive frontiers in hydraulic fracturing, shale gas and rare earth materials) that characterises the bulk of the world economy. Further, we need to consider that, like environmentalism, organised labour is not a homogeneous movement, thus the divisive line between sustainability and business-as-usual crosses through this camp as well. An emerging literature in Environmental Labour Studies[16] shows how the generalised adoption of a green growth/EM discourse on the part of many trade union confederations translates into little effective action because of the internal resistance coming from a number of sectoral unions, as well as contrasting signals coming from employers and governments.[17]

Moreover, although international trade unions confederations have aligned themselves with the hegemonic bloc, a number of organisations, most of them representing unwaged or subsistence workers (such as La Via Campesina, or the Landless Movement of Brazil) have positioned themselves with the counter-hegemonic bloc, and some unions do endorse an EJ or even an anti-capitalist agenda:[18] the most common tendency in this case is that of adopting a social unionism approach, forming large coalitions with environmental and social justice organisations. Beyond all the difficulties typically confronting coalition-building, however, these experiences face the opposition of even potentially progressive governments: this was the case in Spain, where a Climate Coalition (Coalición Clima) was founded in 2008 by 30 ENGOs, trade unions, researchers, consumer organisations and coops, which put forward three draft bills (on mobility, energy and environmental taxing), all rejected by the then socialist government.[19]

Moving from the global to the European scale, however, we can say that most labour organisations, represented by the European Trade Unions Confederation, are unequivocally aligning themselves with a 'green growth' bloc, advocating for a strategy of Just Transition based on public investments and tax incentives for creating blue-collar jobs in the 'clean energy', transport and construction sectors.[20] Initiatives such as the Climate Jobs campaign[21] represent the most advanced version of this position, in the sense that they aim to overcome internal divisions within the labour front and build larger social coalitions to actively mobilise for the adoption of a green jobs (or climate jobs) agenda. This strategic positioning has certainly to do with the fact that labour is currently facing the most serious and enduring economic crisis of the last decades, thus the green/climate jobs perspective appears as the one most likely to spur social and political consensus.

The problem with the Just Transition strategy, as anticipated in Chapter 5, is that it is locking up labour even more firmly with the continuation of capitalism and wage labour in the 'green' mode – built upon the perpetuation of a gendered and racialised division of labour on the world scale – while ruling out a serious discussion of different perspectives and more radical alternatives, and thus the possibility to effectively eradicate the structural causes of both ecological and social inequalities. In this respect, labour's eco-modernism presents a striking similarity with the case of the New Urban Agenda highlighted by Kaika. In both cases, the endorsement of EM on the part of political and trade union leaders rests upon a disregard for those dissenting voices and alternative praxes that locate themselves outside of (green) capitalism, claiming the social pre-eminence of reproduction, use value and the commons – in short, of meta-industrial labour. For a critique of labour's eco-modernism, the key questions then become: What alternate ideas and praxes were historically embedded in labour environmentalism? How did labour movements come to embrace EM?

To answer these questions, a good starting point is that of considering the intellectual roots of Marxist political ecology in Western Europe, identifying the internal tensions and contradictions that derive from divergent readings of Marx and Engels, as much as from Marx's own work.[22] In this vast theoretical debate, I will single out here two points that seem particularly relevant to a critique of the political ecology of labour in Western Europe. The first concerns the very notion of work in the (eco) socialist society: according to J.B. Foster, the Marxist tradition contains two different visions, one centred on the social potential of technology, scientific organisation and automation, leading to the progressive reduction of working time and the increasing of leisure time; the other centres on the dealienation of labour and the consequent liberation of its full potentialities for human development. To Foster, the second vision alone – which he traces to William Morris and to Marx – would be conducive to 'a genuine sustainable society'.[23] He criticises the first vision, which he traces to the US writer Edward Bellamy, for 'taking production as economically and technologically determined' and focusing instead on leisure as a greater social end: a tendency that he sees represented today by the degrowth movement, via the influence of André Gorz.

The second point of the eco-Marxist debate that is of particular relevance here regards the idea, advanced by Foster and Clark in a previous essay, that European Marxists have been mainly influenced by the Frankfurt School's critique of the 'domination of nature', a philosophical anti-Enlightenment stance that, though grounded in writings from the early Marx, had 'led to the estrangement of thinkers in this tradition not only from the later Marx, but also from natural science – and hence nature itself'.[24] According to the authors, this explains why Western Marxism was ill equipped to respond to the rise of the environmental movement in the 1960s, and remained so until a 'second stage ecosocialism' arose in the late 1990s, based on a new reading of ecology as a scientific concept inherent to Marx's thought.

In aiming to explain the convergence of labour movements with EM, this chapter departs from the above two points in various respects. Focusing on the last quarter of the twentieth century, a period in which the influence of Western Marxism on labour movements was still strong, although declining, I show how the distinction between the two visions of work highlighted by Foster had become increasingly blurred, being challenged by various inputs coming from the changing political and economic scenario. Central to these changes, I argue, was the question of political agency and subjectivity: in other words, the question of what role the labour movement (and the working class in general) should have in the ecological revolution. I locate Gorz' 'liberation from work' perspective within this political scenario, connecting it with contemporary political ecology visions developed in Italy and the UK. Regarding the second point, I show how the distinction between Western Marxism and environmentalism was also quite blurred, as ecological concerns were being elaborated, in more or less direct reference to Marx, from within the labour movement itself. I will thus build an alternative narrative, seeking to trace the connections between labour and environmental concerns as they were being reformulated via a dialectical process that involved tensions and convergences between Marxism (or rather, various forms of Western Marxism) and a variety of political movements across the Left spectrum, namely: labour, environmentalism (especially anti-nuclear movements), feminism and degrowth.

The chapter's narrative goes as follows: by the mid-1970s, an early version of Marxist political ecology had taken shape in Italy, which found coherent expression in the work of the communist politician and intellectual Laura Conti. Based on an understanding of ecology as the science of biophysical interrelations, and on a vision of ecosocialism as science-based planning, this political ecology envisioned a strong urban/industrial working class as the only political subject capable of leading the rest of society towards the ecosocialist horizon. In this sense, it can be considered an early expression of labour's ecological modernism, which rested upon the progressive power of the productive forces, understood as male blue-collar workers in heavy industry and infrastructures. In the same years, a different political ecology perspective was coming from the Austrian-French intellectual André Gorz: also based on a scientific understanding of ecology, this position differed from the previous insofar as it reflected contemporary discontentment with traditional ideas of working-class subjectivity and the ideology of work, while also breaking the nexus between the development of productive forces (or economic growth) and socialist ecological planning. In Gorz, the revolutionary ecological subject ceases to be the male blue-collar worker to become an undefined, multitude-like social subject who refuses to be identified on a class basis. This position reflected an incipient crisis of labour movements, consequent to a complex restructuring and tertiarisation of labour markets in the core Western European economies between the 1970s and 1980s. Together with the emergence of a strong green movement in those same countries, agglutinated by a common anti-nuclear stance (a position that labour movements were not ready to take), the 1980s marked a divergence between labour and environmental movements, and the entrenchment of the former on the defence of economic growth at any cost. This increasing divergence and opposition formed the political trend to which a leading British Marxist intellectual like Raymond Williams felt the need to respond, offering his own early Marxian (but also Polanyian) understanding of livelihood, or reproduction, as the common ground between labour and environmental concerns. Marking a definite departure from earlier visions of ecology as science-based planning, Williams exhorted the labour movement to move away from the production ideology that had held it hostage to the development of capitalist productive forces. Unlike Gorz, however, he maintained that the political subject for the ecological revolution could only be the labour movement – once this had elaborated the necessary 'qualitative alteration of socialism'.

Summing up the narrative so far: while Conti's ecosocialism was to be built at the point of production, Gorz and Williams pointed towards reproduction (or livelihood) as the terrain where a different political ecology of labour could have been built. For this to become a hegemonic vision of labour environmentalism, however, something was missing: reproduction needed to be more clearly theorised as the bearer of political agency and subjectivity. This was a step that the feminist movement had taken in the previous decade, without being able to carry out the necessary triangulation with labour and environmentalism. The German scholar/activist Maria Mies exemplifies this point: moving from a critique of the colonial and sexual divisions of labour that underlie capitalist accumulation, Mies had elaborated a rethinking of the labour/ecology nexus based on the social centrality of reproductive work. These ideas, developed in a period of great turmoil for European socialism due to the fall of the Soviet bloc, were foundational for the development of materialist ecofeminist thought, but did not influence the evolution of labour environmentalism. By the turn of the twentieth century, in fact, most of trade union organisations and Left parties in Western Europe had steered away from ecosocialism and were officially embracing the mainstream EM perspective. The disconnect between labour and ecofeminist movements, I argue in the Conclusions, prevented the former from completing the 'qualitative alteration of socialism' that it needed to address the ecological crisis on its own terms, without succumbing to the logic of green capitalism. Opening to the material ecofeminist understanding of work and political subjectivity might have led Western labour movements towards a convergence with the global EJ perspective instead.

The next section identifies Italy as the place where an early political ecology of labour took shape from within the Marxist tradition in the mid-1970s, and then evolved into a mass social movement, rescinding its original nexus with the labour movement a decade later.

THE 'ECOLOGY OF CLASS': LAURA CONTI AND THE ITALIAN LEFT

Since the early 1970s, the thesis of ecological crisis as a contradiction of capitalism had appeared in Italy via what was then the biggest communist party in Western Europe, the PCI. During its

1971 cadres' school in Frattocchie (notably, one year before the ecological crisis was officially declared by the Club of Rome), the PCI had held its first national meeting on the theme 'Man (sic), nature, society', where party executive Giovanni Berlinguer[25] had admitted the need to update Marxist orthodoxy in order to take into account the concept of natural limits. He had compared ecology to socialist planning and emphasised the need for the party to consider the environment a working-class priority.[26] A few years later, the public intellectual and PCI representative Laura Conti published her *Che cos'è l'ecologia. Capitale, lavoro, ambiente* [What Is Ecology. Capital, Labour and the Environment] (1977), a book which offered a theorisation of this new vision and the elaboration of a corresponding political strategy. A physician by training, and a passionate science writer, Conti became a leading figure in the birth of an Italian left-green movement.[27] She defined political ecology as the study of how social relationships within the human species influence the natural world and other species, and described ecology as a metabolic relationship between society and the environment. The key thesis in the book was that such a complex web of interrelationships between natural and social mechanisms needed a good dose of environmental planning, to be democratically designed and governed. The struggle against those who damage nature, 'the life of our and other species', Conti wrote, could not be left in the hands of the market or some eco-technocracy, but must have society as a protagonist, and specifically one social class: the one that opposed capital. In defending not only its own interests, but those of humanity itself as belonging to the sphere of nature, the working class would find substantial solidarities and coalitions in society, Conti believed.

The relationship of labour with environmental movements was a rather marginal topic in Conti's approach: she took note in passing of the antipathy between (industrial) workers and environmentalists, but this to her simply represented a healthy reaction of the working classes towards a certain type of (middle-class) environmentalism based on moralistic condemnations of mass consumption and on the defence of some charismatic mammal.[28] Rather than on an unlikely alliance with this kind of environmentalism, Conti's political ecology was philosophically grounded on Marx's concept of alienation from species-being. For her, this pointed to the need for labour organisations to develop their own ecological revolutionary reformism, that is, an 'ecology of class', in order to counter capitalism's abuse of the environment from within.[29]

The 'ecology of class', however, was ridden by an apparently irreducible internal contradiction, of which Conti was critically aware, what I would call the ecomodernist dilemma of labour: the advancement of a working-class ecological consciousness was consequential to the development of the forces of production, in the sense that only an advanced industrial apparatus could grant the occupational levels and political strength that were necessary for workers to develop their role as defenders of the environment. In Conti's vision, labour environmentalism had to work with this contradiction: she thought that a possible response was for the Left to push towards the development of those industrial sectors that granted the higher occupational level with the lower relative environmental impact. Even though only limited and temporary solutions such as this could be found, she noted, this was a struggle that the labour movement could not postpone.

Conti's response to the ecological crisis was somehow representative of the official line adopted by the communist party in this respect – or at least, that which was most acceptable within the eurocommunist politics of the late 1970s and early 1980s: rather than rejecting industrial modernity in its entirety, the communists had to exercise their power within and outside the Italian parliament for keeping the destructive power of the forces of production at bay, via democratic control. This democratisation of production started at the workplace, where a new methodology of participative workers' control had been elaborated in a collaboration between the confederate unions and a handful of militant scientists, and reached society at large via public institutions which the workers' movement was struggling for painstakingly in those same years, such as the national public health system.[30]

The 'ecology of class' approach, however, presented other limitations, both internal and external. To begin with, the strategy of workers' control over industrial toxicity via trade unions and public institutions was highly vulnerable to cycles of economic recession, especially in industrial production, which restricted the possibilities for capital to invest in health and safety measures and impaired the negotiating power of unions in the workplace. The strategy was also blind to geographical, sectoral and gender differentiations within the Italian working class: it not only proved unsuccessful in the 'underdeveloped' south, but it also reinforced an implicit acceptance of male blue-collar work in heavy industry as the only meaningful form of political mobilisation in an area of social policy – that of environmental and public health – which entailed much broader significance and subjectivities.

Nevertheless, this approach did produce important results in terms of convergence between red and green politics, among which the founding of the Italian League for the Environment (Legambiente), today a well-established Italian ENGO, created in 1979 as a subsection of the PCI recreational branch. Conti played a leading role in the foundation of the new organisation, together with a handful of public intellectuals who shared a common militancy in the PCI.[31]

During the 1980s and 1990s, the 'ecology of class' strategy was thoroughly challenged by the changing structure of the Italian economy, with the tertiarisation and then precarisation of the labour force, and environmental concerns gradually shifted from the point of production to that of consumption.[32] Consequently, Legambiente started to detach itself from the PCI by refusing centralised strategies and promoting instead forms of territorial organising 'outside the workplace, and away from any logic of economic defence'.[33] The definite detachment came in 1986, together with a choice for privileged relationships with the Green Party in local and national elections. The crisis of the 'ecology of class' approach became clear when the Left (the communist and socialist parties and the CGIL union) split along two opposite fronts facing the anti-nuclear referendum promoted by the Greens and environmental organisations in 1987.[34]

Legambiente's move from the factory to the territory thus reflected a process of historical division within the Italian Left, more than between labour and environmental organisations, a division that became more evident with the split of the PCI into two political formations in 1991. This coincided with a marked shift in the eco-Marxist debate that found expression in the magazine *Ecologia Politica–Capitalismo Natura Socialismo* (CNS),[35] founded in 1991. The magazine introduced the thesis of ecological crisis as capital's 'second contradiction', put forth by US scholar James O'Connor in his introduction to the first issue of CNS,[36] while also popularising different versions of radical ecological thinking.[37]

In short, rather than representing a shift from the materialist to the post-materialist approach to environmentalism – where materialism is reductively intended as 'economic defence' – Italian political ecologists expressed a more specific shift: from the political hegemony of the PCI, based on the ideological hegemony of the urban industrial proletariat, to a grassroots, territorial perspective on ecology, where a diversity of social subjectivities were to be aggregated towards the defence of the conditions of production against industrial toxicity. This shift, however, came after the concept itself of working class had been subjected to a radical political ecology critique. Let us now turn to examine that critique.

FAREWELL TO WORKING-CLASS ECOLOGY: ANDRÉ GORZ AND THE 'LIBERATION FROM WORK'

In 1977, the same year in which Conti's *What Is Ecology* came out in Italy, the notion of the ecological crisis being a political issue of special relevance for socialism was being raised, on radically different grounds, by Austrian-French intellectual André Gorz in his essay 'Freedom and Ecology',[38] first published in France as *Écologie et Liberté* (1977). Inspired by a kind of humanist socialism deriving from a variety of non-orthodox Marxist influences (from Sartre's existentialism and Marcuse's critique of domination, to the May 1968 students' movement and the Italian so-called autonomous Marxism), Gorz introduced the theme of nature as an external limit to growth and defined political ecology as the struggle for a democratic and emancipatory politics of the environment. The essay presented the ecological crisis as a crisis of reproduction by overaccumulation, and exposed the risks of economic reductivism deriving to labour from an uncritical adoption of the productivist ethos – a thesis he developed in full length in his later *Critique of Economic Reason* (first published as *Métamorphoses du travail*,

1988).[39] Gorz' political ecology was thus intended not only as a critique of the capitalist mode of production, but also as an instrument of liberation from the conceptual categories that constrained the labour movement within the capitalist order: a political project 'capable of, at the same time, overcoming capitalism and transforming socialism itself'.[40] Also in later writings, he maintained a constant effort at dialoguing with various labour organisations, and with trade unions in particular, inviting them to challenge the 'ideology of work'.[41]

Gorz did not see the overcoming of capitalist industrialism as a return to the pre-industrial order of medieval craftsmanship and village economy. Rather, he developed a socialist utopia where people regained control over industrial technology, not via centralised forms of state planning but via individual and community self-management. If ecology was not compatible with capitalist rationality, it was not compatible with authoritarian socialism either. Moreover, like ecology, so technology was not socially neutral: only those technologies that were compatible with capitalism (or with state socialism) in their drive for the quantitative growth of production – regardless of qualitative considerations – had been developed thus far. The clearest example being that of nuclear power, presupposing a de facto authoritarian social order. Consequently, the strive for a different society passed through the struggle for different technologies: 'socialism is no better than capitalism if it does not change tools', he stated.[42] Only those technologies that could be controlled at the community scale, bring about greater individual or local autonomy, preserve the reproduction of life, and facilitate producers' and consumers' control should be developed.[43]

Gorz' political ecology was an intellectual's call to the labour movement for the transformation of socialism. Making a much greater impact, however, was Gorz' by far most popular book, *Farewell to the Working Class*, significantly subtitled *An Essay on Postindustrial Socialism*,[44] first published in France as *Adieux au Prolétariat* (1980) and two years later translated into English and Italian. The book presented the thesis that the time had come for the labour movement to emancipate itself from the ideology (and ethic) of work-as-employment, and move towards a post-industrial revolution which fully acknowledged the new historical subject formed by the 'non-class of postindustrial proletarians'.[45] Published on the verge of deindustrialisation and post-Fordist restructuring of production in Western Europe, the book envisioned the end of the age of full employment and Keynesian welfare, to be replaced by a society of 'freed time' based on the autonomous production of use-value – a thesis that sounded then fresh and timely. The core idea it presented was, however, very old and even foundational to Marxian thought: 'Communism is neither full employment nor a wage for everyone' – Gorz wrote – 'but the elimination of work in the socially and historically specific form it has in capitalism. That is to say, it is the elimination of work-as-employment, work-as-commodity.'[46] This was a mostly forgotten dimension of Marxist thought that Gorz had derived from non-orthodox Marxists such as the East German Rudolf Bahro and the Italian Antonio Negri.[47]

The key point in the book was that capitalism would not end by its own internal contradictions, nor by reaching its external limits – namely, the ecological. On the contrary, he wrote, in the past 20 years capitalism had demonstrated an unexpected ability 'to manage the non-resolution of its problems, accommodate its dysfunctions, even drawing renewed strength from this state of affairs'. Furthermore, such problems would remain unresolved even if the working class took control over the forces of production – that is, without changing them qualitatively. The sad news, in fact, was that capitalism had produced a working class whose immediate interests were more consonant with the reproduction of capitalism itself than with 'a socialist rationality'. Consequently, the eradication of capitalism could only come from those areas of society that embodied or prefigured 'the dissolution of all social classes, including the working class itself'.[48] Gorz' suggestion was not that of replacing the traditional Marxist working class with a different but equally transcendent subject, a new class with a new mission of historical salvation. Like all nascent social movements, he wrote, that of 'those who refuse to be nothing but workers' was a non-class with a strong liberation ethic, 'a negation of law and order, power and authority, in the name of the inalienable right to control one's life'.[49] The movement was thus not structured as such, because its main concern was the building of individuals' autonomy. This was, at the same time, its main strength and weakness, 'because spaces of autonomy captured from the existing social order will be marginalized, subordinated or ghettoized unless there is a full transformation and reconstruction of society'.[50]

As Martin Ryle and Kate Soper have noted, André Gorz gave voice to an alternative, radical thought on ecology and labour that was emerging between the 1970s and 1980s, one which posited a new connection 'between the human desirability of a less work-dominated life and the environmental necessity of a less indiscriminately productivist economy'.[51] In Gorz' *Écologie et liberté*, and in some intellectual circles in France, this idea started to be called '*decroissance*' (degrowth), a concept that has received increasing attention in the wake of the current economic crisis, especially in Southern Europe.[52] This renewed interest has given birth to what now considers itself an umbrella movement – a convergence of social movements and intellectuals that strive for decoupling social welfare and the good life from the imperative of economic growth.[53] Gorz' ideas on the 'liberation from work' are central to today's degrowth movement. Needless to say, however, the demise of the Fordist organisation of work and the abandonment of a working-class-centred politics have not resulted in the desired 'liberation from work' – quite the opposite, one could say. There was in Gorz no further elaboration on which alternative possibilities could be developed in replacement of the old class perspective, nor a critical analysis of the 'new social movements' that had occupied the scene of ecological activism, mostly via anti-nuclear mobilisation, evolving into new political formations – the green parties.

To sum up, Gorz gave voice to a diffused disillusionment with the opportunity, and even the possibility to develop an 'ecology of class' strategy, based on the political subjectivity of the labour movement. This disillusionment located itself within a historical phase of post-industrial transition experienced by France, Italy, Germany and most of Western Europe in the 1980s, which, coupled with an almost generalised shift to neoliberal policies of 'flexibilisation' of work and welfare cuts, contributed to a serious weakening and crisis of the labour movement. At the same time, the post-industrial transition seemed to open the way to a 'liberation from work', or 'refusal of work', as theorised in the Italian workerist/Autonomia movement: a position that was radically alternative to, and also vehemently opposed by, the traditional labour movement.

The crisis of the 'ecology of class' approach signalled a situation in which political ecology could not be seen any longer as a territory hegemonised by the labour movement: the environmental movement was consolidating itself as a 'new social movement' grown out of radical, grassroots mass mobilisations against nuclear power all over Western Europe,[54] so that it was not possible for Marxists and the traditional Left to depict it as an expression of elitist concerns with charismatic wildlife, as Conti had done before. In other words, the 1980s were a time when it became evident that a red-green politics needed to be built upon a new relationship between labour and environmental movements. The next section will thus illustrate how this possible alliance was envisioned by one of the most prominent Marxist intellectuals of the time: the British critic Raymond Williams.

'A QUALITATIVE ALTERATION OF SOCIALISM': RAYMOND WILLIAMS ON LABOUR AND ECOLOGY

In June of 1984 Williams was invited by the Socialist Environment & Resources Association of Letchworth, UK, to deliver a speech on 'Ecology & the Labour Movement'. He started with the claim that: 'No political development is now more necessary than a convergence between the ecology movement and the labour movement.'[55] Offering a compelling narrative of how work and nature related to each other under industrial capitalism, he envisioned a possible convergence between labour and environmental movements, based on two prerequisites: first, that labour be willing to replace the notion of 'production' with that of 'livelihood'; second, that the environmental movement recognised capitalism as the enemy of nature.

Williams characterised ecosocialist politics in the following terms: the ecological crisis was a product not of modernity in itself – intended as the ability to feed more people out of a limited amount of resources, the ability to escape the Malthusian trap – but of capitalist modernity. The latter was to be understood as a mode of production in which both labour and the environment were considered 'raw material' (or 'resources') for accumulation and profit, rather than an end in themselves. The point for the labour movement was to change that system, not to run it more efficiently. Williams' political ecology converged with Gorz' in rejecting a vision of systemic change that implied a return to pre-industrial agriculture and craftsmanship – he rejected the idea that these modes of production would be able to maintain the current levels of population in Britain. A non-capitalist modernity seems to be the horizon towards which he thought ecology and labour could merge.

In Williams' account, the (UK) labour movement had emerged from a specific environment, that of the first industrial revolution, characterised by the enclosure of common lands, and consequent essential changes in food production and livelihoods; and new ways of 'drawing power from nature' related to the coal-and-iron technology. All this gave society an unprecedented capacity to transform nature, a transformation which became a common work experience. 'Out of that chaotic and dynamic experience, a labour movement was formed, which was primarily defensive', he noted, for it formed around the notion of remediable poverty. And the answer to poverty that was on offer by the social order, he noted, was: 'produce more and you will no longer be poor-work-harder'. Socialists soon realised that there is no necessary correlation between increased production and the reduction of poverty, because the social relationships set up in the course of production determine in large part the decisions about how the surplus is distributed. This makes the possibility for great poverty to continue amidst great wealth – if this is functional to the social order. Thus, the problem was not production per se, said Williams, but the relations of production. For the majority of its history, he claimed, the labour movement had not taken this difference seriously.

The problem Williams was pointing to was labour's entrenchment within the dominant instrumental rationality of capitalist modernity, which labour, he seemed to say, had accepted as an ineluctable fact of life. However, he noted, with the evolution of industrial capitalism, raw materials had become redundant, and population had reached a point much higher than the demand for labour, so that people had become redundant as well, like a raw material that could be left in the ground. And at this point, Williams claimed, the labour movement faced its greatest crisis, a crisis of ideas:

> because, if it continues to see more production as the way to fight poverty, it is simply going to produce its own general redundancy. It is a process which has a certain iron law about it ... So the battle now is ... whether you will continue to accept the predominant mode of production ... or whether you can (and this is where the convergence with the ecology movement must happen) begin to think of a different social order.

The time had come, in other words, to recognise that the problematic relationship of labour with ecology derived from the passive acceptance of this misleading habit of thought, by which production was the necessary and sufficient answer to poverty, while wage labour and private property had become undisputed social institutions. Breaking through this political direction implied abandoning the idea that socialism could compete with capitalism in running the same system better, that is, 'producing more', because the long-term consequences of this model on people and the environment had become unmistakeably clear.

What was the alternative to that? To 'challenge the notion of production as it has been presented, and start with a different conception', Williams stated, one around which the convergence between the socialist and ecology movements could take place, that is, the idea of livelihood. Livelihood was to Williams 'a much deeper concept, and much more human concept than production', because the latter was 'nearly always a gross quantity, indiscriminate as to what you produce, what quality you produce, what effects that production has on others and other kinds', and this prevented the system from prioritising real human needs. On the contrary, livelihood meant 'starting from human place, and from the interest of all living beings involved'. Williams' formulation of the relationship between ecology and labour in terms of livelihood seems consistent with O'Connor's theory of ecological Marxism, formulated a few years later,[56] and signals a tendency of Marxist political ecologists towards converging with the Polanyian critique of commodification that was becoming popular in leftist circles of the time. The originality of Williams' approach, however, consisted in wondering how could livelihood be sustainable in a modern, industrial economy: this was, to him, the core dilemma in the relationship between ecology and labour. The answer that the environmental movement had given was not satisfactory, he claimed, and that was the main reason why the labour movement needed to step in and take the matter into its hands. While questioning the priorities of the production system, in fact, the environmental movement had done so by calling the system soft names – industrial society or modern society – and thus it had never got to 'the hard political choices'. If environmentalists could get to the point where they identified capitalism as the enemy of nature, and if – at the same time – the labour movement were prepared to move in the same direction, then a common basis for red-green politics could be found. If that conversation succeeded, it would produce not simply a convergence between two movements, he concluded, but 'a qualitative alteration of socialism'. And the only force which could carry through this transformation was 'the force which is rooted in the majority interest and in the indispensable livelihood of all the people in the society, and that … ideally … is the labour movement'.[57] To sum up, Williams shared Conti's idea that a strong labour movement is the social subject capable of leading an ecological revolution, but, like Gorz, he did not consider this as a necessary and natural political choice for the labour movement, at least not until a convergence had been realised between this and the environmental movement on the terrain of a politics of livelihood. While all three shared a Marxist critique of capitalism's ecological contradictions, they lacked an understanding of coloniality/racism and patriarchy/sexism as fundamentally constitutive of industrial capitalism itself (and, to a large extent, of then existing state socialism as well), and thus of the capitalist world-ecology.[58] Such awareness was being produced by the ecological strand of the feminist movement, to which the next section is dedicated.

REDEFINING LABOUR: MARIA MIES AND THE MATERIALIST ECOFEMINIST APPROACH

Since the early 1970s, a much needed reflection on domestic and reproduction work as a crucial 'hidden abode' of capitalism had emerged in Italy, France, Germany and other Western countries,[59] contributing in fundamental ways to destructuring old conceptions of work centred on the industrial workforce and on waged employment in general. What is less well known is the fact that this reflection allowed a number of scholars and activists to also develop a materialist ecofeminist perspective based on a critique of the sexual division of labour on the global scale. Probably the most important contribution in this respect came from the German academic and long-time feminist Maria Mies, whose *Patriarchy and Accumulation on the World Scale*, first published in Germany in 1986, soon became a key reference in materialist ecofeminism.[60] The relevance of the book to a discussion of labour and ecology should not be underestimated. It pointed to the same problem that Williams had identified in his Letchworth speech, that of abandoning a reductive understanding of production as the only terrain where poverty could be addressed – but it did so by adding two perspectives that were utterly absent from Williams (as well as in Conti and Gorz): that of women's hidden work and that of the international division of labour. The book focused on what Mies called 'the general production of life, or subsistence production', which she saw as 'mainly performed through the non-wage labour of women and other non-wage labourers as slaves, contract workers and peasants in the colonies' and that for her constituted 'the perennial basis upon which "capitalist productive labour" can be built up and exploited'.[61]

The book added a fresh perspective on the sexual division of labour, theorising it as an issue that went beyond the family sphere, and that defined an entire social structure, including both class relationships and social relations with nature. In a decisively non-essentialist fashion, Mies explained men/women differentiations as evolved out of a historical process, thus criticising the naturalisation of (women's) reproductive labour in Marx and Engels, for whom reproductive activities 'do not belong to the realm of 'productive forces', of 'labour', 'industry and exchange but to 'nature'.[62]

'By separating the production of new life from the production of the daily requirements through labour, by elevating the latter to the realm of history and humanity and by calling the first "natural", the second "social"', she wrote, 'they have involuntarily contributed to the biological determinism which we still suffer today. With regard to women and their labour, they remain as idealistic as the German ideologues whom they criticised.'[63]

To Mies, Marx and Engels' vision ultimately reflected the historical process by which the patriarchal order had been incorporated by European feudalism first (via the witch hunt) and then by industrial capitalism, when 'the sphere where labour power was reproduced, the house and the family' was redefined as 'private, domesticated nature', while the factory was defined as 'the place for public, social ("human") production'.[64] Even though this distinction belongs more fully to the housewifisation process of the Fordist period than to the early industrial era – when a great number of factory workers were women – it remains of great relevance in understanding the social subordination of women as typically non-political subjects, and of reproduction as a non-political terrain of social life.

In the following two decades, Mies gave pathbreaking contributions to the global ecofeminist literature with the books *Ecofeminism*[65] and *The Subsistence Perspective*,[66] which established strong connections between ecology, feminism and the emerging anti-globalisation (or alterglobalisation) movement. They contributed substantially to the discussion of work beyond wage labour and the urban-industrial proletariat, and showed how the possibility for developing a 'good life' beyond capitalist growth was already practised in a number of rural and developing nation contexts. Having emerged from within a scholarly/activist group in 'women and development' studies across Germany, Austria and the Netherlands,[67] the 'subsistence perspective' was thus an empowering concept that gave value to people's abilities to cooperate with each other and with nature in the 'production of life'.

Mies' work resonated with that conducted in the same period by Italian autonomous Marxist feminists Silvia Federici and Mariarosa Dalla Costa, and by the British ecosocialist scholar and activist Mary Mellor.[68] What these arguments had in common was they went beyond the traditional, typically male-dominated national labour movements, to look at exploitation and solidarities at the world scale, where the majority of working-class people were rural women living in developing nations.[69] The global/sexual division of labour was thus the level at which these authors and activists, many of whom came from a Marxist feminist background, developed an ecological critique of capitalism from the point of view of reproduction and livelihood.[70] This intellectual/activist work has been built through a materialist approach to both ecology and feminism, but also in a transnational conversation that intersected in important ways with post-colonial and Third World studies, while operating a difficult dialogue between constructivist and realist approaches to 'nature'.[71] The common analytical terrain was that of exposing the material effects of the capitalist/patriarchal/colonial order in terms of depletion of both ecosystems and people, via extraction of unpaid labour from (mostly women) reproducers and from nature, which originated in the cultural depreciation of reproductive services, while leading to their commodification.

Ecofeminist political economy is of fundamental importance to a consideration of ecology and labour insofar as it exposes the dangers of Western cultural dualisms (between 'culture' as mind-agency-production-masculine and 'nature' as body-passivity-reproduction-feminine) that are deeply enmeshed within socialist politics as well. Its central concern is the need for thoroughly revisiting the notion of what counts as 'labour'. However, the materialist ecofeminist debate took place mostly in 'global' settings such as the first Rio Earth Summit, or 'women and development' conferences,[72] crucially involving women scholars and activists working in the Global South; additionally, most ecofeminist scholars and activists developed their work outside Europe. As a consequence, the impact of this debate on the development of a red-green political agenda in the old continent was very limited, if not absent. The materialist ecofeminist perspective was born and developed as an outsider to the traditional labour movement, and has remained so to this day. Marginalised, and also misunderstood by so-called third wave white feminism,[73] it has gained new momentum in the last few years due to a materialist turn in gender studies, but especially thanks to the growing mobilisations of Indigenous, peasant, and other racialised women against the increasing commodification and depletion of life in the new millennium.

CONCLUSIONS

The last three decades of the twentieth century have represented a crucial turning point in labour environmentalism. The approach that had been consolidated in the Fordist era, based on trade unions' struggles for health and safety regulations at the point of production, extending to society at large via democratic planning – what Italian communists called 'the ecology of class' – lost its centrality, and a variety of different visions emerged on the European Left. None of them, however, succeeded in preventing the labour movements of Western Europe from losing their anti-capitalist perspective and from embracing an ecomodernist political ecology. This defeat should be read against the historical background of structural and political constraints (economic stagnation, deindustrialisation and the end of the Soviet experience); however, it must be also explained as the effect of internal shortcomings of Marxist political ecology: namely, the disconnect between ecosocialist and ecofeminist visions.

At the time of writing, the prospects for a red-green politics in Europe seem to be now polarised around two blocs, which could be seen as broadly representative of, respectively, EM and EJ: the first bloc revolves around a labour-friendly green growth plan based on a mix of market and non-market regulation, as represented by the ETUC and the social democrat group at the European Parliament; the second bloc is inspired by a post-development and global EJ vision, as represented by the degrowth movement, towards which the materialist ecofeminist perspective has converged, and which also incorporates a reclaiming of the 'common' as the political terrain for (ecological) anti-capitalist politics.

The first option represents the official position of labour environmentalism: however, this is now understood in a quite different way than its ecosocialist version. The JT and climate jobs strategies, in fact, see workers not as the political subject of an ecological revolution, but as potential victims of climate policies. In other words, whereas the ecology of class was a transformative strategy, oriented towards a class-based defence of reproduction, this new version of labour's ecomodernism is a conservative strategy, built around the defence of production. The second option, on the other hand, manifests in grassroots resistance to both carbon-intensive activities and 'clean energy' megaprojects, as well as in a number of urban squatting/gardening/work-sharing initiatives, many of which are consciously adopting degrowth principles. According to degrowth advocate and scholar Giorgos Kallis, these actions are not inspired by an escapist, but rather by a nowtopia attitude, that is, one that aims at changing the city by linking grassroots with institutional action.[74] It must be noted, however, that this strategy has failed so far to gain mass traction with the impoverished and precarised working classes of the austerity era, nor does it seem capable of having a constructive dialogue with the labour movement in general.

The disconnect and even occasional hostility between an ecofeminist perspective, now strictly allied with EJ/post-development/degrowth movements, and labour's ecomodernism is what is currently impairing the possibility of developing a stronger and more convincing anti-capitalist ecological struggle, both in Europe and at the global level. This strategy, I contend, should aim at transforming labour environmentalism into an anti-patriarchal and anti-colonial alliance between industrial and meta-industrial workers. For this to happen, a new generation of political ecologists and militant scholars will need to take up the challenge of rethinking the working classes and their ecological agency.

7
The Labour(s) of Degrowth*

Intervening in a debate between Giorgios Kallis and John Bellamy Foster,[1] this chapter engages with the centrality of work and class in the transition to a post-carbon and post-capitalist paradigm. It argues that there can be no degrowth without the dealienation of labour.[2]

Kallis and Foster agree on advocating for an ecological socialism that might be able to democratically regulate the much needed decrease in social metabolism: the central issue being how to carry out the social transformations that will lead to the desired result. While Foster emphasises the need for a new 'ecological revolution' inspired by *The Communist Manifesto* and by a historical materialist approach to earth system science, Kallis asks which institutions will allow a democratic control over social metabolism to be better realised. Both authors also put forward a list of radical policy proposals that they consider achievable under the present conditions and necessary to 'mobilize the general public' (in Foster's words). I believe that what is missing to move forward with this common plan is a clearer vision of what political subjects and which processes of political subjectivation can make it happen. In other words, rather than presupposing a 'general public' as the recipient of any political strategy, we need to build such strategy upon a more solid analysis of the social forces involved, their mutual relations and their possible common interests. In what follows I will offer my contribution in this sense by reflecting on the place that labour movements and working-class people can and should have in degrowth politics and in the transformation of social

---

* This chapter is a slight revision of Stefania Barca, 'The Labor(s) of Degrowth', *Capitalism Nature Socialism* 30, no. 2 (2019): 207–16. Reprinted with permission from Taylor and Francis.

metabolism more broadly. Kallis' main argument that growth of biophysical throughput is still possible in a non-capitalist or even socialist economy is a useful starting point. The argument touches upon an important issue, of interest to all those who connect ecological struggles to an anti-capitalist perspective (from Naomi Klein to ecosocialists). It is reinforced by the observation that, for the most part, socialist regimes have shown levels of environmental devastation fundamentally similar to those of the capitalist world. In this sense, some authors have come to argue that, rather than Capitalocene, the Anthropocene should actually be renamed as Growthocene.[3] Ecological critiques notwithstanding (materialist or otherwise), and especially with the development of nuclear power and synthetic chemicals in the post-Second World War era, both systems followed the imperative of economic growth, which can be seen as the leading cause of ecological unsustainability and 'environmental violence'.[4] If degrowth ultimately means eliminating the productive reinvestment of surplus value,[5] the problem arises of who decides on how the surplus should be dispensed with (*dépense*) and how. This is clearly a political problem. And, I would add, it is one shared by all growth-oriented societies, both capitalist and centrally planned, because in both systems the producers (however defined) are typically estranged from decisions over the allocation of the surplus. In this sense, the emphasis that Kallis' commentary puts on workers as decision makers on surplus allocation sounds misplaced. In state socialism, decisions on what to do with the surplus have been alienated from workers as much as (if not more than) in capitalism. In capitalist systems, the surplus tends to be reinvested in increased production (but also spent in conspicuous consumption, charity, control of the media, etc.). In neither case do workers have much democratic control over the allocation of the surplus produced through their labour.

This is why the dealienation of labour is relevant to degrowth politics. As Leigh Brownhill, Terisa Turner and Wahu Kaara explained in their contribution to a degrowth symposium hosted by the journal *Capitalism Nature Socialism* in 2012, dealienation is the process by which Marx's four forms of estrangement – from the products of labour and the natural world, from the labour process, from species-being and from other humans – are actively reversed through collective action. In their words, 'De-alienation is about action by the exploited and dispossessed, both waged and unwaged', aimed at un-enclosing resources and establishing new commoning practices and social relations.[6] Now, what exactly this entails is highly contingent on local situations and political choices regarding the extent to which the industrial division of labour will be accepted in the future degrowth society. Andre Gorz, a widely recognised inspiration for degrowthers, famously claimed that a certain grade of alienation from the labour process was inevitable in industrial societies.[7] Nevertheless, my point is that degrowth should aim for a truly democratic, worker-controlled production system where this alienation is actively countered by a collective reappropriation of the products of labour and by a truly democratic decision-making process over the use of the surplus. Estrangement from the products of labour is the specific aspect of alienation that concerns me here, insofar as it relates to the separation between the producers and the allocation of surplus that has historically characterised both capitalist and socialist regimes. My hypothesis is that the alienation of the producers from the products of their work is what leads to the reinvestment of surplus into increased production. Consequently, the project of building a degrowth society can only start from fostering dealienation by reopening the possibility for workers' control and economic democracy, from the workplace to society at large.

This, I argue, is the reason why the degrowth movement must build a constructive dialogue with the alienated and exploited workers of the world. Here, in the messy reality of everyday re/productive work, complex contradictions arise that need to be addressed in fundamentally new ways. Different forms of metabolism clash with each other and produce environmental conflicts, which enter into communities' and people's lives, questioning identities, crushing certain life forms and turning them into cogs of the dominant social metabolism.[8] This process takes different forms in the different but interconnected spaces of the global political economy. The fundamental political problem for the degrowth movement, I believe, is to gain a clear perspective on how the alienation of workers occurs, and how it can be reversed.

A good vantage point from which the contradictions in social metabolism can be analysed is the perspective of those workers whose livelihood depends on fossil-driven economic growth, and whose voice rarely makes it to degrowthers' ears. At the time when I was writing this chapter, one of those voices reached me from a distant place through an article published on The Leap website.[9] It was that of a Mapuche oil worker from Patagonia, who told a sad story of dispossession and destruction of local agriculture by Argentina's powerful oil business sector. Together with a history of racial discrimination and state repression, this left him and thousands of others no choice but to join the extractive industries. Working in the oil fields for 25 years, he came to know firsthand the devastating impact they had on his community's land and bodies, and he lost two family members to cancer due to the widespread contamination of water in the area. Despite all this, he considered himself to be fortunate in having a job that allowed him to pay his bills and medical expenses, and to buy bottled water – especially when comparing his situation to that of local farmers who are literally on the verge of starvation. This gives us a measure of how difficult it is for many workers to even consider the possibility of losing their job, no matter how dirty and dangerous, in the absence of viable alternatives.

In this Mapuche man's experience, environmental violence was inextricably linked to alienation from the labour process: once hired, he reported, workers in the oil industry are made to sign a confidentiality agreement 'that gives away (their) right to speak out publicly'; in addition, they are trained in what the company calls 'environmental safety', which means that whatever disaster may happen, the blame is immediately shifted onto their supposed errors. The truth is, however, that disasters occur almost invariably (in this and many other cases) because management orders workers to keep production going despite reported faults or potential leaks. And, if they question choices internally, they face a variety of repercussions. This tells us that weak unions and virtually non-existent enforcement of labour regulations play a major role in determining the environmental impact of production.

Nevertheless, this worker was perfectly aware of the root causes of this situation, and of the negative balance left behind by promises of prosperity based on oil extraction. His quest was one for dealienation, in the sense of gaining control over not only the labour process and product, but also the political process where decisions are made over the best route to prosperity for his country. As he claimed, 'We, as people, have to question and ask ourselves: what gives us more prosperity?' The answer for him was in the development of a flourishing and diversified agriculture without oil, based on the rich natural resources of the country, rather than on soy and other soil-eroding monocrop plantations. This kind of sustainable agricultural development, he explained, is what would give people the opportunity to flourish by getting back 'what's theirs' – that is, the product of their labour. Following his vision, we might imagine, if not degrowth, at least a prosperous way out of fossil-driven economic growth, built upon a dealienated relationship of workers with the labour process and the product of their labour.

The centrality of dealienation in a discussion of degrowth becomes even clearer when we analyse the concrete historical examples in which dealienated workers have been able to enact sustainable modes of production, that is, of working-class environmentalism. While the history of twentieth-century environmentalism is ridden with conflicts between environmental activists and workers, which have compromised any possibility for political alliance in many cases, it also shows important – if less well-known – stories of labour environmentalism, some of them opening the possibility for truly emancipative ways of organising social metabolism. Probably the most well-known example is that of the rubber tappers' struggles of the 1980s, which initiated the emancipative conservation experience of the Amazon 'extractive reserves'.[10] But other stories can be dug out of oblivion, and other voices from working-class and labour environmentalism can be heard. One example is that of the occupied factory Ri-Maflow in Italy, a former producer of auto components, which went bankrupt and laid off 320 workers in 2009. After the new owners had dismantled and taken away all machinery, a group of former workers organised a coop and occupied the space, with the idea of reappropriating it as a starting point for building new forms of production, consumption and waste disposal. Adopting the slogan 'Re-use, re-cycle, re-appropriate', these workers have initiated a workshop of computer and appliance repair, a flea market, and the processing and distribution of local produce.[11] They also run the place as a space for community music and arts activities and social events, and for hospitality to refugees. Their plan was to collect enough resources to be able to turn these and other activities into a stable form of income, that is, to develop dealienated forms of work and production.

The labour/degrowth relationship is complicated by the fact that, even when they claim to be sensitive to climate and environmental issues, trade unions and labour parties in the capitalist world are mostly locked in the growth paradigm, rather than in an anti-capitalist perspective of dealienation. Their proposals for Green New Deal or Just Transition get trapped into the idea that a green capitalism is possible, by which they mean a set of public policies can be implemented that would reduce carbon emissions while stimulating the green economy and creating 'decent' jobs. In other words, the majority of trade unions now aim for positive changes that would address the multiple current crises of ecology, economy and social inequalities without waiting for some systemic change that is difficult to envision and agree upon.[12] This perspective is not to be dismissed too easily, as it does represent the official position of large labour confederations and so-called blue–green alliances, which have the possibility to orient union policies at the national and local level, and might influence public investment choices among alternative options, for example, between coal and solar power. The Just Transition proposal, for example, is premised upon the notion that the shift to a post-carbon economy will inevitably imply massive layoffs of workers who are dependent on the fossil economy, and thus consequent suffering in their communities. Degrowth cannot avoid considering this aspect of the transition to a non-fossil-based and substantially different production system. Therefore, degrowth policy proposals must include concrete recommendations for dealing with those foreseen layoffs, sustaining the livelihoods of working-class communities in the transition process, and replacing fossil-generated wealth with different forms of income and welfare. All this compels us to engage with the discussions and positioning that are expressed by organised labour at different levels. Simply dismissing organised labour as a non-relevant actor in the transition to a post-carbon, post-capitalist or degrowth society will not do.

When taking into consideration the workers' perspective, we also need to be aware of the limited extent to which labour organisations represent the global working class, and of the differentiations and fractures that cross the non-homogeneous world of both organised and non-organised labour. One clear example is given by the Keystone XL pipeline controversy in the United States, where five national trade unions expressed vocal support, two openly declared their opposition to it (both representing domestic workers, mostly women with an immigrant status), while the remaining unions adopted a 'none-of-our-business' attitude. This caused extreme uneasiness and contrasting attitudes among the base (union locals and individual workers), for it often posed them on the opposite side of the struggles conducted by their communities against the pipeline.[13] Similar examples of other configurations of positions assumed by organised labour in face of climate politics could be drawn, for example, the One Million Climate Jobs campaign in South Africa. Here, the exacerbation of both economic and climate inequalities has led to an alliance between parts of the labour movement with environmental justice and green movements. This alliance has been able to reclaim the Just Transition strategy, filling it with radical anti-capitalist and anti-neoliberal content. From a simple claim for jobs in a green growth agenda, the campaign has moved to advocate for public disinvestment from the fossil and nuclear sectors, coupled with energy democracy and food sovereignty at the community level.[14]

What these cases exemplify is the fact that there exist, at this historical conjuncture, concrete possibilities for articulating degrowth and labour politics in new ways, via grassroots mobilisations in community unionism and social movement unionism, pushing labour organisations towards a radical critique of the growth paradigm.[15] This articulation is a crucial starting point for developing new forms of political-ecological consciousness that go beyond the current divisions between organised labour and degrowth and environmental justice movements. In fact, they would allow for the development of an emancipatory ecological class consciousness, premised upon a reconceptualisation of work in the sense of dealienation and commoning, which is a necessary prerequisite for a socially just degrowth strategy. The articulation of degrowth and labour politics towards an ecological class consciousness implies the important task of reconsidering ideas of class in general, and of working class in particular. Class is still an important reality for most of the world population, even as it intersects with multiple other social differentiations. Simply ignoring class politics will do a disservice to degrowth: for example, it will obscure the (largely) white middle-class nature of the movement and thus its possibilities for political action.

In order to reconsider class and its intersection with degrowth, a good starting point is enlarging the concept of class relations beyond the wage labour relation and towards a broader conception of work as a mediator of social metabolism. Biophysical throughput is largely the result of work done in the factory, the field, the office, the retail centre and the household, but workers have very limited control over the process. In capitalist societies, wage relations and growth-oriented state politics alienate workers from both the labour process and product. As ecofeminist political economy has abundantly shown, production takes over and dominates reproduction, the surplus is accumulated or reinvested in the infinite expansion of the system. This has negative consequences for the quality and quantity of resources available for the reproduction of life and for the entire biosphere. Logically speaking, working-class people – those who are located at the bottom of the global labour hierarchy and who pay the higher price to its social costs – have a vested interest in the subversion of this system. This is what I call an emancipatory ecological class consciousness: the awareness that climate change (and environmental violence in general) is the newest form of class war – as always, articulated with gender and racial domination – and that it needs to be combated via struggles for dealienation and commoning.[16] Being locked in the growth society, however, working-class people have a limited ability to make sense of and struggle against the current organisation of social metabolism, as the Mapuche oil worker's testimony makes clear. That is why contributing to the emergence of ecological class consciousness is, I argue, a crucial task for a degrowth movement that rejects authoritarian solutions to the ecological crisis and aims at building large social alliances around a truly emancipatory transition to a post-carbon economy.

In embracing the challenge of raising ecological class consciousness, degrowthers can count on the fundamental support of feminist praxis. Feminist political economy, especially in its articulations with ecofeminism and feminist political ecology, has offered invaluable contributions for a rethinking of work, and intersects in new ways with the degrowth debate.[17] Through dialogue with post-colonial studies, this literature has produced a thorough critique of GDP and development politics as inextricably linked to undervaluation of subsistence economies and reproductive (mostly women's) work, what Ariel Salleh has named 'meta-industrial labor'.[18] From a Marxist perspective, the work of reproduction is mostly (even if not entirely) carried out outside of wage relations, but is inextricably linked to them via the constant need for capital to appropriate this work in order to sustain production and accumulation.[19] Ecofeminist scholars see the ecological crisis as a global manifestation of the gendered division of labour, and thus a major cause of the crisis of social reproduction.[20] What ecofeminist and feminist political economists have in common is the identification of reproduction as a crucial terrain for anti-capitalist struggle and ecological revolution.[21] Overall, what ecofeminist political economy tells us is that combating the capital/state appropriation of the reproductive and care labour carried out by the global meta-industrial working class is a crucial step towards the dealienation of the labour process and towards taking control over the surplus in a global commoning perspective. In this sense, the degrowth movement should listen to the voices coming from the emancipatory ecological class consciousness that is already guiding the struggles of many women-led movements on the margins of the global political economy.[22]

Ultimately, my argument here boils down to a single overarching question: 'What is the political subject of a degrowth revolution?' I think this subject should not be confined to an ecologically minded global middle class willing to reduce consumerism and work addiction, and/or to engage in direct action to express its disappointment with economic/environmental policies. This has historically played an important vanguard role in raising consciousness of the degrowth perspective and in allying with a variety of other movements and ideas. But this approach will remain politically weak unless it manages to enter into dialogue with a broadly defined global working class – including both wage labour and the myriad forms of work that support it – and its organisations. This is a very difficult endeavour in the present conditions, but it might become a concrete possibility if we accept the implications of degrowth for the meta-industrial workers of the world and if the degrowth debate takes on a clearer direction towards emancipation from both the alienation of wage labour and the capitalist (or state) appropriation of reproductive labour. This, I believe, is what is needed in order to decrease social metabolism while increasing social wellbeing and equality.

Epilogue
Care Work in the Post-Carbon Transition: Reflections on the Global Climate Jobs Campaign*

This book has made a case for the inclusion of care and subsistence workers – reproductive labour – into a political and analytical redefinition of labour; at the same time, it has shown the central role played by reproductive workers in environmental struggles throughout the Great Acceleration era – a centrality that has not been fully acknowledged so far by the trade union and labour movement. Based upon these lessons learned from environmental history and political ecology research, and on my own involvement with climate justice campaigning internationally, here I offer some reflections on their implications for reformulating the politics of post-carbon transition.

Over the past decade, climate justice organising has offered a unique opportunity for overcoming the historical fracture between labour and environmental politics at the grassroots level, showing that many on both sides are unconvinced by the jobs versus environment discourse, and understand the planetary crisis as resulting from, and further deepening, social inequalities on all scales. This is no trivial political achievement, signalling how a new era is opening up for grassroots organising towards systemic change. Part of this new wave of climate activism, the global Climate Jobs Campaign (CJC) is an international network of grassroots organisations in various countries battling around national plans for

* This chapter is based on my intervention at the workshop 'International Labour movements in the struggle for a Just Transition – policy proposals and strategies' (with live stream), part of the Global Climate Jobs Conference held in Amsterdam on 6–8 October 2023.

a post-carbon transition centred on public investments in carbon-reducing jobs and infrastructures.[1] Carried out via both direct actions and grassroots counter-planning, the CJC is based on the idea – solidly backed by data – that 'key sectors that have a direct impact on emissions', such as 'energy, transport, construction, forest management and agriculture', need to be thoroughly restructured in order to substantially reduce their carbon footprint, and this restructuring offers a unique opportunity for creating new jobs.[2] These sectors, in fact, are seen as essential to a just post-carbon transition, that is, one that does not imply impoverishment and deprivation of resources for the population. For the largest part, the envisioned 'climate jobs' would provide the work needed to build a new energy and transport infrastructure, and retrofit buildings for energy efficiency; in addition, they would be used in forest management for fire prevention and in carbon-neutral farming. Also, the CJC calls for substantial expansion of public transport, in order to make sure that the shift to electric mobility is accompanied by a reduction of per-capita energy consumption; and for the new jobs to be in public employment through publicly funded projects, so as to make sure that the transition is made in the interest of workers and the climate, rather than private profit.

I believe this is a radical-realist approach that offers the historic opportunity to respond to the climate crisis from below, that is, by bringing together labour and ecological movements on a common post-carbon transition platform. The inclusion of forest management and agriculture in a climate jobs plan is also a very important innovation, especially in the European context, and a welcome acknowledgment of the relevance of environmental care to climate mitigation.

Nevertheless, the CJC also reflects several of the limitations and dilemmas which are being debated within the global climate justice movement, as well as in activist-research circles. This is because old patterns are hard to die, and tend to remain active in people's minds and organisational structures, limiting the potential of the new. A major problem, for example, seems to be an excessive reliance on the state as a guarantor of public interest, overlooking the important limitations to national sovereignty due to the current international trade regime, as well as the financial constraints that limit public funding in poorer countries. Second, the emphasis on the shift to renewable energies seems to disregard the huge environmental impacts of large-scale renewables, which imply inevitable tradeoffs in terms of land and water use. A relevant example is given by the transport sector: electric vehicles and high-speed trains imply lithium extraction and the substantial landscape changes associated with railway constructions. Such costs tend to be borne by rural communities, contributing to the destruction of more sustainable forms of production and living, and are being in fact opposed by local movements all over the world.[3]

These limitations suggest that jobs creation is a necessary but insufficient way to ensure a just post-carbon transition; issues of overall consumption reduction (or degrowth), of energy democracy, of territorial sovereignty, and much more complicate the 'climate jobs' strategy, but also need to be part of the conversation. What I am interested in discussing here, however, is one aspect of the CJC which seems to be significantly limiting its political potential: the tendency to disregard the relevance of reproductive work – both waged and unwaged – in the post-carbon transition. Among the number of new jobs that are envisioned by the CJC, none is foreseen for the care sector, be it in domestic, public or community spheres; nor for subsistence food provisioning, be it in peasant or urban farming, non-industrial fishing or agroforestry. This leaves subsistence and care workers – once again – out of the conversation, disregarding what they have to say, and missing out on their unique potential for tackling the planetary crisis.

I believe this shortcoming of the CJC may be explained with a tendency to inherit the underestimation of reproductive work which has historically characterised the labour movement.[4] Like the trade union discourse on Just Transition, so the CJC is based on the assumption that the two major crises of our time are the climate crisis and the unemployment crisis. For a long time now, however, feminist movements and studies have been arguing that we live in the midst of a care crisis with no historical precedent. This is because the neoliberal turn of the last three decades has produced a serious lack of investment in care at the public policy level, with enormous social costs and an exponential increase in the burden of unpaid care, most of which continues to be carried out by economically and socially marginalised women. Indisputable data has been produced over the last four decades about the immense amount of unpaid working hours that are necessary for the re/production of human beings and for the satisfaction of their material and immaterial needs. Non-remuneration affects caregivers unequally, depending on social class, citizenship, skin colour, gender, sexual orientation, degree of specialisation and access to resources. These inequalities make care work an area of immense, although largely invisible, social injustice.[5]

At the same time, the fiscal crisis in the public sector has led to the intensification of extractivism, with the consequent increase in pollution, environmental risk and diseases associated with mining and other projects. Although not easily quantifiable, this increase in environmental risks generates an additional unpaid and invisible workload, due to the need to take care of the bodies and territories affected. Furthermore, not only the climate crisis and environmental pollution, but the biodiversity crisis and the loss of non-human life have taken on genocidal dimensions, with dire consequences for the very evolution and survival of the human species – including the zoonoses that gave rise to the current coronavirus pandemic.

On the other hand, as this book has repeatedly shown, the 'forces of reproduction' have been contrasting environmental degradation with a constant work of protection, maintenance and regeneration of reproduction conditions. Much of this work is unpaid, while being systematically made invisible and silenced in the public sphere.

In rural settings, a huge amount of unpaid care work concerns subsistence agriculture and territorial integrity: soil quality and regeneration, water conservation, seed saving and sharing, landscape and agroecosystem maintenance, preservation and socialisation of local knowledge – all of this must be recognised as essential work with great ecological relevance. In many parts of Europe, territorial care has long been practised by the elderly peasant population, but since the 2008 financial crisis, younger generations have increasingly been demanding their right to a dignified life in the countryside. Being mostly invisible and unappreciated, this work is constantly threatened by, but also resistant to, extractive industries and financial powers that intend to appropriate what they see as natural resources. Throughout the European periphery, for example, governments currently look to territories as resources that can offer revenue to pay off the national debt. As a consequence, many hours of individual and collective work are being used to organise anti-extractive resistance, that is, to *keep gas under the grass*, and *oil in the soil*, as well as to develop, support and fight for regenerative agriculture and alternative development projects.

In urban environments, a great deal of unpaid (self-)care work is carried out by individuals and organised groups in response to the unmet needs of impoverished people, dilapidated public housing, abandoned industrial zones and crumbling urban centres. Urban farming practices offer local, cheap and fresh food, simultaneously producing green space and oxygen while regenerating the soil and stopping cementation. Waste pickers and secondhand markets contribute to low-impact reuse and recycling practices. Street artists and activist groups revitalise abandoned spaces, creating opportunities for non-commodified socialisation, as well as economies of solidarity and mutual aid. Community volunteers help people survive and access essential services. Although this work is invisible and not counted in the national fiscal balance, it is essential for the wellbeing of all those who do not have access to a middle-class lifestyle. It also helps resist the expansion of alienated, commodity-based relationships between people and with the city, bringing neighbourhoods together and generating communities. In times of climate or public health disasters (like the one we are experiencing) these networks activate solidarity and mutual support activities that are essential for survival, recovery and restoration.

In short, in both urban and rural environments, an enormous amount of work is spent on daily reproductive activities that aim to guarantee the wellbeing of people and territories, compensating for the lack of adequate public services and the exhausting effects of capitalist production. From a feminist perspective, this work should be properly counted and recognised as emission-reducing. By connecting the climate and care crises, both before and after the Covid-19 pandemic, feminist activists and theorists have been claiming that care work is relevant to the politics of Just Transition because it holds high social value while being low in carbon emissions; yet, care work is seriously overburdened and underfunded everywhere, but especially in financially weak and indebted economies. Massive public investments in care jobs are thus needed to respond to this contradiction. According to the Platform for a Feminist Green New Deal, for example, care is valuable, low-carbon, community-based work that should be re-evaluated and central to our new economy.[6] Similarly, the Blueprint for Europe's Just Transition (launched in 2019) considers care 'for people and the planet' as an integral part of its 'green public employment' programme, and proposes the implementation of a 'care income'.[7] In response to the care crisis highlighted by the pandemic, the care income principle has then been propagated by the Global Women's Strike movement through an international campaign that combines the fight for counting and remunerating informal care work with a new perception of the relevance of this work for the health of people and the planet.[8] This proposition is based on counting and valuing the hidden contribution of reproductive work to the reduction of carbon emissions and, more broadly, to human and non-human wellbeing. The claim of the Care Income campaign is that such recognition would lead to adequately remunerate such essential work.

Following up on these claims from the eco/feminist movements, I propose, the CJC could move towards a truly radical-realist approach by demanding a significant amount of new employment in reproductive work. This could be done via a plan for creating public employment in caregiving, local food provisioning and urban commoning.

To begin with, a huge amount of 'climate jobs' could be created by simply compensating unpaid caregiving: this would reduce the total amount of hours spent in paid work, which constitutes a primary cause of CO_2 emissions, while relieving caregivers from the double burden of paid and unpaid work. At the same time, a substantial expansion and restructuring of healthcare services via proximity-based and community-run infrastructures would reduce the huge carbon footprint of centralised healthcare; and substantial investments in schooling, educational and recreational opportunities, including collective school transport and local food provisioning, would improve living conditions in rural areas and urban peripheries. Further, 'climate jobs' might include financial compensation and training for people to reduce the carbon footprint of their own homes, or of the homes and institutions where they work, by devising and practising sustainable forms of housekeeping.

In the food sector, 'climate jobs' might take the form of publicly funded and community-run agroecology, agroforestry, non-industrial fishing and other forms of subsistence food provisioning for local markets, to be developed in both rural and urban settings. These would significantly reduce the carbon content of food provisioning, with important reflections on the preservation of local habitats, knowledge, cultivars, seeds, biodiversity, air and water quality, and generational renewal in farming, while also aiming at food sovereignty. Access to such projects should be granted to anyone who demands it, including immigrant populations, allowing for rural revitalisation and cultural diversity.

Further, the CJC could include a programme of jobs creation in urban community care via substantial and non-aleatory funding for urban commons, so that they become permanently available to those in need, while maintaining their grassroots and self-organised character. This would turn cities into spaces which are liveable for all, by granting that buildings, parks and infrastructures are renewed and sustainably managed, minimising their environmental impact and in the interest of those who live there; that individual consumption and waste are substantially reduced and circular economies are autonomously created; that green space is expanded in the place of new constructions.

Summing up, considering caregiving as essential and low-carbon work constitutes a unique opportunity to create new 'climate jobs' with their own potential to reduce carbon emissions, prevent disasters, safeguard socioecological wellbeing and increase equality and social inclusion. Such a truly radical Just Transition could unite different kinds of socially useful and emission-reducing work on a common political platform. The material basis for such political unity is given by the fact that all working-class people, of all genders and sexes, urban and rural, have the same material needs: access to cheap and safe food, energy, housing, education, healthcare, and a healthy environment for all present and future generations. By including reproductive labour in its plans to create new jobs, the CJC could widen the sphere of potential alliances to sectors and social movements that have remained external until now – thus overcoming the fictitious separation not only between the labour and climate movements, but also between the productive and reproductive sphere, as well as between urban and rural movements and communities. By facilitating the unity of all these potentially transformative forces, the campaign would thus become a stronger and truly radical-realist platform for climate justice.

Notes

INTRODUCTION

1. GA research, led by the late Will Steffen, was part of a broader scientific project, initiated by Paul Crutzen at the International Geosphere Biosphere Program (IGBP), which aimed at quantifying human impact on earth as far back as possible. One major result of that research programme was the proposed definition of a new geological epoch termed Anthropocene. Even though no consensus has been reached to date on whether this is appropriate, and when this new epoch would start, many agree that the post-1945 era would be the most likely start, since all the available data point towards a sharp increase in human impact upon terrestrial ecosystems – hence the Great Acceleration term. In 2016, the Anthropocene Working Group of the International Union of Geologic Sciences (IUGS) recommended that the year 1950 serve as the starting point for identifying the new epoch, based on the possibility of studying the fallout of plutonium isotopes, caused by nuclear weapons testing, as an observable signal in rock strata. See Will Steffen, Wendy Broadgate, Lisa Deutsch, Owen Gaffney, and Cornelia Ludwig, 'The Trajectory of the Anthropocene: The Great Acceleration', *The Anthropocene Review* 2, no. 1 (1 April 2015), https://doi.org/10.1177/2053019614564785; John P. Rafferty, 'Anthropocene Epoch', in *Encyclopedia Britannica* (12 September 2023), www.britannica.com/science/Anthropocene-Epoch.
2. Steffen et al., 'The Trajectory of the Anthropocene'.
3. I have used the expression 'master's narrative' to signify a representation of human history which is shaped by and uncritically reflects colonial, class, sex/gender and species inequalities: see Stefania Barca, *Forces of Reproduction: Notes for a Counter-Hegemonic Anthropocene*, 1st edn (Cambridge University Press, 2020), https://doi.org/10.1017/9781108878371.
4. See Stefania Barca, 'Telling the Right Story: Environmental Violence and Liberation Narratives', *Environment and History* 20, no. 4 (1 November 2014), https://doi.org/10.3197/096734014X14091313617325.
5. In their 2015 version, the socioeconomic data for all trends except two (primary energy use and international tourism) was disaggregated in three blocks: OECD countries, BRICS (Brazil, Russia, India, China and South Africa), and Others. This allowed us to get some sense of the huge inequalities in the respective levels of wealth, and consequently of the respective responsibilities for the planetary crisis. See 'Great Acceleration', International Geosphere Biosphere Programme (IGBP), accessed 15 September 2023, www.igbp.net/globalchange/greatacceleration.4.1b8ae20512db692f2a680001630.html.

6. J.R. McNeill and Peter Engelke, *The Great Acceleration: An Environmental History of the Anthropocene since 1945* (Cambridge, MA: Harvard University Press, 2014).
7. International Labour Office, a tripartite United Nations agency that represents trade union, business and governmental actors with a focus on labour-related policies.
8. Raj Patel and Jason W. Moore, *A History of the World in Seven Cheap Things: A Guide to Capitalism, Nature, and the Future of the Planet* (2017; reprinted, Oakland: University of California Press, 2018).
9. Giovanna Di Chiro, 'Living Environmentalisms: Coalition Politics, Social Reproduction, and Environmental Justice', *Environmental Politics* 17, no. 2 (April 2008): 276–98, https://doi.org/10.1080/09644010801936230; Carolyn Merchant, *Earthcare: Women and the Environment* (London: Routledge, 1996); Ariel Salleh, *Eco-Sufficiency & Global Justice: Women Write Political Ecology* (London: Pluto Press, 2009); Stacy Alaimo, *Bodily Natures: Science, Environment, and the Material Self* (Bloomington: Indiana University Press, 2010); Greta Gaard, 'Feminism and Environmental Justice', in *The Routledge Handbook of Environmental Justice*, ed. Ryan Holifield, Jayajit Chakraborty, and Gordon Walker, 1st edn (New York: Routledge, 2017), 74–88, https://doi.org/10.4324/9781315678986-7; Sherilyn MacGregor, ed., *Routledge Handbook of Gender and Environment*, Routledge International Handbooks (London and New York: Routledge, Taylor & Francis Group, 2017); Celene Krauss, 'Women and Toxic Waste Protests: Race, Class and Gender as Resources of Resistance', *Qualitative Sociology* 16, no. 3 (1993): 247–62, https://doi.org/10.1007/BF00990101; Dianne Rocheleau, Barbara Thomas-Slayter, and Esther Wangari, eds, *Feminist Political Ecology: Global Issues and Local Experiences*, International Studies of Women and Place (London: Routledge, 1996).
10. Ariel Salleh, *Ecofeminism as Politics: Nature, Marx and the Postmodern*, 2nd edn (1997; reprinted, London: Zed Books, 2017); Marilyn Waring, *Counting for Nothing: What Men Value and What Women Are Worth* (Toronto: University of Toronto Press, 1999); Rocheleau, Thomas-Slayter, and Wangari, *Feminist Political Ecology*.
11. Enara Echart and Maria del Carmen Villarreal, 'Women's Struggles against Extractivism in Latin America and the Caribbean', *Contexto Internacional* 41, no. 2 (August 2019): 303–25, https://doi.org/10.1590/s0102-8529.2019410200004; Angelika Sjöstedt Landén and Marianna Fotaki, 'Gender and Struggles for Equality in Mining Resistance Movements: Performing Critique against Neoliberal Capitalism in Sweden and Greece', *Social Inclusion* 6, no. 4 (22 November 2018): 25, https://doi.org/10.17645/si.v6i4.1548; Francisco Venes, Stefania Barca, and Grettel Navas, 'Not Victims, But Fighters: A Global Overview on Women's Leadership in Anti-Mining Struggles', *Journal of Political Ecology* 30, no. 1 (21 February 2023), https://doi.org/10.2458/jpe.3054.
12. Merchant, *Earthcare*.
13. This claim is central to the entire research field of feminist economics: see Günseli Berik and Ebru Kongar, *The Routledge Handbook of Feminist Economics* (New York: Routledge, 2021).

14. Barca, *Forces of Reproduction*.
15. See Stefania Barca, 'On Working-Class Environmentalism: A Historical and Transnational Overview', *Interface* 4, no. 2 (2012): 61–80.
16. Giulia Malavasi, *Manfredonia: storia di una catastrofe continuata* (Milano: Jaca book, 2018).
17. I understand anti-master struggles as those which oppose the intersected inequalities of race/coloniality, sex/gender, class and species, and the logic of GDP growth as the uncontestable primary aim of public policies. See Barca, *Forces of Reproduction*.
18. See for example: Silvia Federici and Arlen Austin, *The New York Wages for Housework Committee 1972–1977: History, Theory and Documents* (Brooklyn, NY: Autonomedia, 2017); Selma James, *Sex, Race and Class: The Perspective of Winning a Selection of Writings 1952–2011* (Oakland: PM Press, 2012); Selma James, *Our Time Is Now: Sex, Race, Class, and Caring for People and Planet* (Oakland: PM Press, 2021); Louise Toupin, *Wages for Housework: A History of an International Feminist Movement, 1972–77* (London: Pluto Press, 2018).
19. Susan Ferguson, *Women and Work: Feminism, Labour, and Social Reproduction* (London: Pluto Press, 2020).
20. James, *Our Time Is Now*; see also the website Global Women's Strike, accessed 13 September 2023, https://globalwomenstrike.net/.
21. 'Caring for the land' is now seen as part and parcel of the Care Income campaign; in fact, over the past few years, the movement has been deeply involved with agroecology and with women farmers' mobilisations in India. See 'Caring for the Land', Global Women's Strike, accessed 31 January 2024, https://globalwomenstrike.net/caringfortheland/.
22. 'Care Income Now', Global Women's Strike, accessed 31 January 2024, https://globalwomenstrike.net/care-income-now/.
23. James, *Sex, Race and Class*; James, *Our Time Is Now*.
24. See for example, Margaret Prescod-Roberts, *Black Women: Bringing It All Back Home* (Bristol: Falling Wall Press, 1980); Wilmette Brown, *Black Women and the Peace Movement* (Bristol: Falling Wall Press, 1984) and 'Roots: Black Ghetto Ecology', in *Reclaim the Earth: Women Speak out for Life on Earth*, ed. Leonie Caldecott and Stephanie Leland (London: The Women's Press, 1983), 73–85; Andaye, *The Point Is to Change the World: Selected Writings* (London: Pluto Press, 2020).
25. Historical and current material on the WFH movement can be requested at the Crossroads Women Centre in London, which hosts various organisations that are part of the campaign.
26. On this concept, see also Amaia Orozco, *The Feminist Subversion of the Economy: Contributions for Life against Capital* (Common Notions Press, 2022).
27. On the concept of anti-master struggle, see Barca, *Forces of Reproduction*.
28. Irina Velicu and Stefania Barca, 'The Just Transition and Its Work of Inequality', *Sustainability: Science, Practice and Policy* 16, no. 1 (10 December 2020): 263–73, https://doi.org/10.1080/15487733.2020.1814585; Stefania Barca, 'Labor in the Age of Climate Change', *Jacobin* (18 March 2016), accessed 31

January 2024, https://jacobin.com/2016/03/climate-labor-just-transition-green-jobs/.

29. Dimitris Stevis, 'Just Transitions: Promise and Contestation', in *Elements in Earth System Governance* (April 2023), https://doi.org/10.1017/97811089 36569; Dimitris Stevis and Romain Felli, 'Planetary Just Transition? How Inclusive and How Just?' *Earth System Governance*, Exploring Planetary Justice, 6 (December 1, 2020): 100065, https://doi.org/10.1016/j.esg.2020.100065; Sean Sweeney and John Treat, 'Energy Transition: Are We Winning?' (New York: TUED, Trade Unions for Energy Democracy, 24 January 2017), accessed 31 January 2024, https://rosalux.nyc/energy-transition-are-we-winning/. I am grateful to Rocío Hiraldo for her help in compiling this information.
30. Barca, 'Labor in the Age of Climate Change'.

1. LABOURING THE EARTH

1. On the Ilva story, see Stefania Barca and Emmanuele Leonardi, 'Working-Class Ecology and Union Politics. A Conceptual Topography', *Globalizations* 15, no. 4 (2018): 487–503.
2. See John Bellamy Foster, *Marx's Ecology. Materialism and Nature* (New York: Monthly Review Press, 2000), 141–78; Paul Burkett, *Marx and Nature: A Red and Green Perspective* (New York: St. Martin's Press, 1999), 25–56; James O'Connor, *Natural Causes: Essays in Ecological Marxism* (New York: Guilford Press, 1998), 29–48. See also Howard L. Parsons, *Marx and Engels on Ecology* (Westport and London: Greenwood Press, 1977), 121–9, 136–45, 148.
3. See Carolyn Merchant, ed., *Key Words in Critical Theory: Ecology* (Amherst: Humanity Books, 1996), 1–27. On alienation from nature, see Foster, *Marx's Ecology*, 72–8. See also Richard Peet, Paul Robbins, and Michael Watts, 'Global Nature', in *Global Political Ecology*, ed. Peet, Robbins, and Watts (Abingdon and New York: Routledge, 2011), 13–15. On Marxism and ecology, see also Carolyn Merchant, *Radical Ecology: In Search for a Livable World* (New York: Routledge, 2005), 142– 8, and Ted Benton, ed., *The Greening of Marxism* (New York and London: Guilford Press, 1996); and Kohei Saito, *Karl Marx's Ecosocialism: Capital, Nature, and the Unfinished Critique of Political Economy* (New York: Monthly Review Press, 2017). See also Chapter 6 in this book.
4. See Raymond Williams, 'Ideas of Nature', in *Problems in Materialism and Culture: Selected Essays*, ed. Williams (London: Verso, 1980), 67–85. On the relevance of this essay for environmental history, see William Cronon, 'The Densest, Richest, Most Suggestive 19 Pages I Know', *Environmental History* 10 (2005): 679–81. On Raymond Williams, see also Chapter 6 in this book.
5. See O'Connor, *Natural Causes*, 48–71.
6. See Richard White, '"Are You an Environmentalist or Do You Work for a Living?" Work and Nature', in *Uncommon Ground: Rethinking the Human Place in Nature*, ed. William Cronon (New York: W.W. Norton, 1996),

171–85; and Richard White, *The Organic Machine. The Remaking of the Columbia River* (New York: Hill & Wang, 1995). On nature as Eden, see Carolyn Merchant, *Reinventing Eden: The Fate of Nature in Western Culture* (New York: Routledge, 2003).

7. Piero Bevilacqua, *Tra natura e storia. Ambiente, economie, risorse in Italia* (Rome: Donzelli, 1996).
8. Emilio Sereni, *Storia del paesaggio agrario italiano* (Rome-Bari: Laterza, 1961; English edition: History of the Italian Agricultural Landscape, trans. with an introduction by R. Burr Litchfield (Princeton: Princeton University Press, 1997). On Sereni and his legacy in the Italian environmental history, see also Piero Bevilacqua, 'Una scelta di campo. Dialogo intorno alla storia del paesaggio agrario italiano', *Zapruder. Rivista di Storia della Conflittualità Sociale* 24 (2011): 134–9.
9. Marco Armiero and Marcus Hall, 'Il Belpaese: An Introduction', in *Nature and History in Modern Italy*, ed. Armiero and Hall (Athens: Ohio University Press, 2010), 1–14.
10. Piero Bevilacqua, 'The Distinctive Characters of Italian Environmental History', in ibid., 15–32. On the nineteenth-century idea of *bonifica*, see also Marcus Hall, *Earth Repair: A Transatlantic History of Environmental Restoration* (Charlottesville and London: University of Virginia Press, 2005).
11. For an environmental history of the Pontine marshes, see Roberta Biasillo, *Una Storia Ambientale delle Paludi Pontine. Terracina dall'Unità alla Bonifica Integrale (1871–1928)* (Rome: Viella, 2023). On the forced migration involved in the *bonifica* of the Pontine marshes seen through a family saga, see, for example, Antonio Pennacchi, *Canale Mussolini* (Milan: Arnoldo Mondadori, 2010).
12. Don Mitchell, *The Lie of the Land: Migrant Workers and the California Landscape* (Minneapolis and London: University of Minnesota Press, 1996), and Don Mitchell, *They Saved the Crops: Labor, Landscape and the Struggle over Industrial Farming in Bracero-Era California* (Athens: University of Georgia Press, 2012).
13. Douglas Cazaux Sackman, *Orange Empire: California and the Fruits of Eden* (Berkeley and Los Angeles: University of California Press, 2007), 152. On labour in the agricultural landscape of modern California, see also Linda Nash, *Inescapable Ecologies: A History of Environment, Disease and Knowledge* (Berkeley, Los Angeles, and London: University of California Press, 2006); Chad Montrie, *Making a Living: Work and Environment in the United States* (Chapel Hill: University of North Carolina Press, 2008), 113–28; Donald Worster, *Rivers of Empire: Water, Aridity and the Growth of the American West* (Oxford and New York: Oxford University Press, 1985).
14. On the environmental history of the Amazon region, see José Augusto Pádua, 'Biosfera, História e Conjuntura na Análise da Questão Amazónica', *História, Cîencias, Saúde – Manguinhos* 6 (September 2000): 793–811, and José Augusto Pádua, '"Drawn by Blind Greed": The Historical Origins of Criticism Regarding the Destruction of the Amazon River's Natural Resources', in *Water, Geopolitics and the New World Order*, ed. Terje Tvedt,

Graham Chapman, and Roar Hagen, vol. 3, A History of Water 2 (London: Tauris, 2011), 176–92.

15. Silvino Santos, *No Páis das Amazonas* (1921), Museu da Imagem e do Som do Amazonas (MISAM), Manaus, Brazil. Many thanks to MISAM for sending a DVD copy to the University of Coimbra in March 2012.
16. Merchant, *Reinventing Eden*, 11–64.
17. Felipe Milanez, 'Faroeste caboclo', *Rolling Stone Brazil* (December 2007): 76–83; see also the report of the Pastoral Land Commission on the violence against rural workers in today's Brazilian Amazon: 'Commissão Pastoral da Terra, Diagnóstico sobre as situações de ameaça de morte contra trabalhadores e trabalhadoras rurais do Sul e Suleste do Pará' (2012); many thanks to Felipe Milanez for pointing me to this source.
18. This is what emerged from a report filed by the British diplomat Roger Casementin 1910–11. See Jordan Goodman, *The Devil and Mr. Casement: One Man's Battle for Human Rights in South America's Hearth of Darkness* (New York: Farrar, Straus and Giroux, 2010). See also Roger Casement, *The Amazon Journal of Roger Casement*, ed. Angus Mitchell (London: Anaconda Editions, 1997). I am thankful to Xenia Wilkinson for pointing me to Casement's report.
19. On Santos, see Flávio Araújo Lima Bittencourt, 'Silvino Santos', Biblioteca Virtual das Amazonas, Série Memória, accessed 16 October 2012, www.bv.am.gov.br/portal/conteudo/serie_ memoria/04_silvino.php.
20. Williams, 'Ideas of Nature', 78.
21. See, for example, Donald Worster, *Dust Bowl: The Southern Plains in the 1930s* (Oxford and New York: Oxford University Press, 1979). On a global scale, see also Paul Josephson, *Industrialized Nature: Brute Force Technology and the Transformation of the Natural World* (Washington, DC: Island Press, 2002).
22. Thomas Andrews, *Killing for Coal: America's Deadliest Labor War* (Cambridge, MA and London: Harvard University Press, 2008).
23. Myrna Santiago, *The Ecology of Oil: Environment, Labor, and the Mexican Revolution, 1900–1938* (Cambridge and New York: Cambridge University Press, 2006).
24. Myrna Santiago, 'Work, Home and Natural Environments: Health and Safety in the Mexican Oil Industry, 1900–1938', in *Dangerous Trades: Histories of Industrial Hazards across a Globalizing World*, ed. Christopher Sellers and Joseph Melling (Philadelphia: Temple University Press, 2012), 41.
25. Arthur McEvoy, 'Working Environments: An Ecological Approach to Industrial Health and Safety', *Technology and Culture* 36 (April 1995): S145–73.
26. Christopher Sellers, *Hazards of the Job: From Industrial Hygiene to Environmental Health Science* (Chapel Hill and London: University of North Carolina Press, 1998).
27. See, for example, Peter Bartrip, *The Way from Dusty Death: Turner and Newall and the Regulation of Occupational Health in the British Asbestos Industry, 1890–1970* (London and New York: Athlone Press, 2001); Francesco Carnevale and Davide Baldasseroni, *Mal da lavoro: Storia della salute dei lavoratori* (Rome-Bari: Laterza, 1999); Ray Elling, *The Struggle for Workers'*

Health: A Study of Six Industrialized Countries (New York: Baywood, 1986); Roy Johnston and Arthur McIvor, *Lethal Work: A History of the Asbestos Tragedy in Scotland* (Tuckwell: The Mill House, 2000); Gerald Markowitz and David Rosner, *Deceit and Denial: The Deadly Politics of Industrial Pollution* (Berkeley: University of California Press, 2002); David Rosner and Gerald Markowitz, *Dying for Work: Workers' Safety and Health in Twentieth-Century America* (Bloomington: Indiana University Press, 1986).

28. See, for example, Sellers and Melling, *Dangerous Trades.*
29. Stefania Barca, 'Lavoro, corpo, ambiente. Laura Conti e le origini dell'ecologia politica in Italia', *Ricerche Storiche* 41 (September–December 2011): 541–50. On Laura Conti, see also Chapter 6 in this book.
30. Laura Conti, *Che cos'é l'ecologia. Capitale, lavoro, ambiente* (Milan: Mazzotta,1977): 135–6.
31. Saverio Luzzi, *Il virus del benessere. Ambiente, salute e sviluppo nell'Italia Repubblicana* (Rome-Bari: Laterza, 2009), 100–1; Wilko Graf von Hardenberg and Paolo Pelizzari, 'The Environmental Question, Employment and Development in Italy's Left', *Left History* 2 (2008): 77–104.
32. Roberto Della Seta, *La Difesa dell'Ambiente in Italia* (Milan: Franco Angeli,2000), 46.
33. Luzzi, *Il virus del benessere.*
34. On the Seveso accident and on the subsequent European Union directives, see 'Industrial Safety', accessed 20 October 2012, http://ec.europa.eu/environment/seveso/index.htm.
35. Laura Conti, *Visto da Seveso. L'Evento Straordinario e l'Ordinaria Amministrazione* (Milan: Feltrinelli, 1977), 56.
36. Laura Centemeri, *Ritorno a Seveso. Il Danno Ambientale, il suo Riconoscimento, la sua Riparazione* (Milan: Mondadori, 2006).
37. Virginio Bettini and Barry Commoner, *Ecologia e lotte sociali. Ambiente, popolazione, inquinamento* (Milan: Feltrinelli, 1976), 6.
38. On the abortion debate in Italy as related to the Seveso accident, see Bruno Ziglioli, *La Mina Vagante. Il Disastro di Seveso e la Solidarietà Nazionale* (Milan: Franco Angeli, 2010).
39. Conti, *Visto da Seveso*, 54.
40. Chad Montrie, *To Save the Land and People: A History of Opposition to Coal Strip Mining in Appalachia* (Chapel Hill: University of North Carolina Press, 2002). On the impact of strip mining on local communities, see also the more recent Michele Morrone and Geoffrey L. Buckley, eds, *Mountains of Injustice: Social and Environmental Justice in Appalachia* (Athens: Ohio University Press, 2011), and Shirley Stewart Burns, *Bringing down the Mountains: The Impact of Mountaintop Removal Surface Coal Mining on Southern West Virginia Communities, 1970–2004* (Morgantown: West Virginia University Press, 2007).
41. Gregory Rosenthal, 'Life and Labor in a Seabird Colony: Hawaiian Guano Workers, 1857–70', *Environmental History* 17 (October 2012): 744–82.
42. Stacy Alaimo, *Bodily Natures. Science, Environment and the Material Self* (Bloomington and Indianapolis: Indiana University Press, 2010).

43. Andrew Hurley, *Environmental Inequalities: Race, Class and Environmental Pollution in Gary, Indiana* (Chapel Hill: University of North Carolina Press, 1995). For a review of US literature on class in environmental history, see Chad Montrie, *A People's History of Environmentalism in the United States* (London: Continuum, 2011), 147–57.
44. Hurley, *Environmental Inequalities*; Robert Gottlieb, *Forcing the Spring: The Transformation of the American Environmental Movement* (Washington, DC: Island Press, 1993); Daniel Faber, ed., *The Struggle for Ecological Democracy: Environmental Justice Movements in the United States* (New York: Guilford Press, 1998). For a synthesis on historical relationships between occupational and environmental hazards in the United States, see John Froines, Robert Gottlieb, Maureen Smith, and Pamela Yates, 'Disassociating Toxic Policies: Occupational Risk and Product Hazard', in *Reducing Toxics: A New Approach to Policy and Industrial Decision Making*, ed. Robert Gottlieb (Washington, DC and Covelo: Island Press, 1995), 95–123.
45. Richard W. Judd, *Common Lands, Common People: The Origins of Conservation in Northern New England* (Cambridge, MA: Harvard University Press, 1997), vii.
46. Gunther Peck, 'The Nature of Labor: Fault Lines and Common Ground in Environmental and Labor History', *Environmental History* 11 (2006): 212–38.
47. Lawrence M. Lipin, *Workers and the Wild: Conservation, Consumerism, and Labor in Oregon, 1910–30* (Urbana and Chicago: University of Illinois Press, 2007); Montrie, *Making a Living*.
48. Scott Dewey, 'Working for the Environment: Organized Labor and the Origins of Environmentalism in the United States, 1948–1970', *Environmental History* 1 (1998): 45–63.
49. Gottlieb, *Forcing the Spring*, and Montrie, *Making a Living*, 91–128.
50. Robert Gordon, '"Shell No!" OCAW and the Labor-Environmental Alliance', *Environmental History* 4 (1998): 460–87.
51. Gottlieb, *Forcing the Spring*, 365–6.
52. Brian Obach, *Labor and the Environmental Movement: The Quest for Common Ground* (Cambridge, MA: MIT Press, 2004).
53. See, for example, Patrick Novotny, *Where We Live, Work and Play: The Environmental Justice Movement and the Struggle for a New Environmentalism* (Westport and London: Praeger, 2000), 41–72; see also William Brucher, 'From the Picket Line to the Playground: Labor, Environmental Activism, and the International Paper Strike in Jay, Maine', *Labor History* 52 (Winter 2011): 95–116. For a Brazilian perspective on environmental justice and work, see Henri Acselrad, Selene Herculano, and José Augusto Pádua, eds, *Justiça Ambiental e Cidadania* (Rio de Janeiro: Relume Dumará, 2004), and Marcelo Firpo Porto, 'Saúde do trabalhador e o desafio ambiental: contribuições do enfoque ecossocial, da ecologia política e do movimento pela justiça ambiental', *Ciença e Saúde Coletiva* 10 (2005): 829–39.
54. Robert D. Bullard, *Dumping in Dixie: Race, Class and Environmental Quality* (Boulder: Westview Press, 2000), 86.
55. Laura Pulido, *Environmentalism and Economic Justice: Two Chicano Struggles in the South* (University of Arizona Press, 1996), 3–56.

56. On the 'environmental conflict' concept, see Robert Bullard, *Dumping in Dixie*, 9– 11, and Joan Martinez Alier, *The Environmentalism of the Poor: A Study in Environmental Conflicts and Valuation* (Cheltenham: Edward Elgar, 2002); for an environmental history approach, see Marco Armiero, 'Seeing Like a Protester: Nature, Power and Environmental Struggles', *Left History* 13 (Spring–Summer 2008): 59–76.
57. David Pellow, *Garbage Wars: The Struggle for Environmental Justice in Chicago* (Cambridge, MA and London: MIT Press, 2002), 12.
58. Ibid.,131–60.
59. Mary Allegretti, 'A construção social de políticas públicas. Chico Mendes e o movimento dos seringueiros', *Desenvolvimento e Meio Ambiente* 18 (July–December 2008): 39–59; see also M. Keck, 'Social Equity and Environmental Politics in Brazil: Lessons from the Rubber Tappers of Acre', *Comparative Politics* 27 (July 1995): 409–24; Gomercindo Rodriguez, ed., *Walking the Forest with Chico Mendes: Struggle for Justice in the Amazon* (Austin: University of Texas Press, 2007).
60. Biorn Maybury-Lewis, 'Introduction to the English Edition', in Gomercindo Rodriguez, *Walking the Forest*, 10–11.
61. Gianfranco Bettin, ed., *Petrolkimiko: le Voci e le Storie di un Crimine di Pace* (Milan: Baldini Castoldi Dalai, 1998); Barbara Allen, 'A Tale of Two Lawsuits: Making Policy/Relevant Environmental Health Knowledge in Italian and U.S. Chemical Regions', and Stefania Barca, 'Bread and Poison: The Story of Labor Environmentalism in Italy', in Sellers and Melling, *Dangerous Trades*, 154–67 and 126–39; Giulio Di Luzio, *Fantasmi dell'Enichem* (Milan: Baldini Castoldi Dalai, 2003); David Allen and Laurie Kazan-Allen, *Eternit and the Great Asbestos Trial* (London: IBAS, 2012).
62. On the 'Just Transition', see for example Dimitris Stevis, *Just Transitions Promise and Contestation* (Cambridge University Press, 2023); for a critical perspective, see Irina Velicu and Stefania Barca, 'The Just Transition and Its Work of Inequality', *Sustainability: Science, Practice and Policy* 16 no. 1 (2020): 263–73. See also Chapter 5 in this book.

2. BREAD AND POISON

1. On the relationship between labour and the environment (especially in the United States) see, for example, Richard White, '"Are You an Environmentalist, or Do You Work for a Living?" Work and Nature', in *Uncommon Grounds: Rethinking the Human Place in Nature*, ed. William Cronon (New York: W.W. Norton, 1996), 171–85. See also Scott Dewey, 'Working for the Environment: Organized Labour and the Origins of Environmentalism in the United States, 1948–1970', *Environmental History* l (1998): 45–63; Scott Gordon, '"Shell No!" OCAW and the Labour-Environmental Alliance', *Environmental History* 4 (1998): 460–87; Brian K. Obach, *Labour and the Environmental Movement: The Quest far Common Ground* (Cambridge, MA: MIT Press, 2004); Gunter Peck, 'The Nature of Labour: Fault Lines and Common Ground in Environmental and Labour History', *Environmental History* 2 (2006): 212–38.

2. On the environmental agency of the labour movement in this period of Italian history, see also Chapter 6 in this book.
3. On the topic of 'embodying working-class history', see Ava Baron and Eileen Boris, 'The Body as a Useful Category in Working-Class History', *Labour: Studies in Working-Class History of the Americas* 4, no. 2 (2007): 23–43, and the following debate in the same journal issues. In particular, in this chapter I refer to Baron and Boris' remark on 'how bodies come to "matter" and how they can cause trouble' (p. 26), and to John F. Kasson's call to 'keep our eyes on the "marks of capital" on workers' bodies' (p. 46). See John F. Kasson, 'Follow the Bodies: Commentary on "The Body as a Useful Category in Working-Class History"', *Labour: Studies in Working-Class History of the Americas*, 4, no. 2 (2007): 45–8.
4. See Amalia Signorelli, 'Movimenti di popolazione e trasformazioni culturali', in *Storia dell'Italia repubblicana*, Vol. 4, *La Trasformazione dell'Italia: sviluppo e squilibri* (Torino: Einaudi, 1995), 589–658; Stefano Musso, 'Lavoro e occupazione', in *Guida all'Italia contemporanea, 1861–1997*, ed. Massimo Firpo, Nicola Tranfaglia, and Pier Giorgio Zunino (Milano: Garzanti, 1998), 485–544; Nicola Crepas, 'Industria', in *Guida all'Italia contemporanea, 1861–1997*, ed. Massimo Firpo, Nicola Tranfaglia, and Pier Giorgio Zunino (Milano: Garzanti, 1998), 287–422.
5. See Kitty Calavita, 'Worker Safety, Law and Social Change: The Italian Case', *Law and Society Review* 2 (1986): 189–228; Francesco Carnevale and Alberto Baldasseroni, *Mal da lavoro: Storia della salute dei lavoratori* (Roma-Bari: Laterza, 1999), 147–229; Giovanni Berlinguer, *Storia e politica della salute* (Milano: Franco Angeli, 1991).
6. See Carnevale and Baldasseroni, *Mal da lavoro*, 230–82; Patrizio Tonelli, 'Salute e lavoro', in *il '900: Alcune istruzioni per l'uso*, ed. Luigi Falossi (Firenze: La Giuntina, 2006), 45–65; Patrizio Tonelli, 'La salute non si vende: Ambiente di lavoro e lotte di fabbrica tra anni '60 e '70', in *I due bieni rossi del '900, 1919–1920 e 1968–1969: Studi e interpretazioni a confronto*, ed. Luigi Falossi and Francesco Loreto (Roma: Ediesse, 2007), 341–52.
7. On the 1968 students' movement in Italy, see Giovanni De Luna, *Le ragioni di un decennio. 1969–1979: Militanza, violenza, sconfitta, memoria* (Milano: Feltrinelli, 2008). On the birth of the 'new industrial hygiene' in Italy, see Carnevale and Baldasseroni, *Mal da lavoro*, 238–44. See also CGIL-CISL-UIL Federazione Provinciale di Milano, *Salute e ambiente di lavoro: L'esperienza degli SMAL* (Milano: Mazzotta, 1976), 189–99.
8. CGIL-CISL-UIL, *Salute e ambiente di lavoro*, 12–13.
9. On the Seveso disaster, see Laura Centemeri, *Ritorno a Seveso; Il danno ambientale, il suo riconoscimento, la sua riparazione* (Milano: Mondadori, 2006). See also Saverio Luzzi, *Il virus del benessere: Ambiente, salute, sviluppo nell'Italia repubblicana* (Roma-Bari: Laterza, 2009), 140–55. On Medicina Democratica, see 'Archivio della categoria: Rivista', last modified 15 May 2022, www.medicinademocratica.org.
10. For a discussion of the 'strong objectivity' approach in the study of industrial hazards, see Barbara Allen, *Uneasy Alchemy: Citizens and Experts in Louisiana's Chemical Corridor Disputes* (Cambridge, MA: MIT Press, 2003), 6–7, 118, 140, 149.

11. CGIL-CISL-UIL, *Salute e ambiente di lavoro*, 110.
12. Ibid., 74–85.
13. Ibid., 143–9.
14. Ibid., 189.
15. On the ANIC (later renamed Enichem) in Manfredonia, see Giulia Malavasi, *Manfredonia: Storia di una Catastrofe Continuata* (Milano: Jaca Book, 2018); see also Giulio Di Luzio, *I fantasmi dell'Enichem* (Milano: Baldini Castoldi Dalai, 2003).
16. On the 1976 accident, see Francesco Tomaiuolo, '1976–2006: Trent'anni di arsenico all'enichem di Manfredonia', *I Frutti di Demetra* 12 (2006): 33–41; and Luzzi, *Il virus del benessere*, 152–5.
17. See Di Luzio, *I fantasmi dell'Enichem*.
18. Ibid.
19. On ecofeminism and women's action for environmental justice, see Carolyn Merchant, *Radical Ecology: The Search for a Livable World* (London: Routledge, 1995), 193–222; on feminist science and action research, see also Allen, *Uneasy Alchemy*, 117–50.
20. Maurizio Portaluri, interview by the author, April 2009.
21. Noted in Barbara Allen, 'A Tale of Two Lawsuits: Making Policy-Relevant Environmental Health Knowledge in Italian and U.S. Chemical Regions', in *Dangerous Trade. Histories of Industrial Hazard across a Globalizing World*, ed. Christopher Sellers and Joseph Melling (Philadelphia: Temple University Press, 2011), 154–67.
22. On the encounter between Lovecchio and Portaluri, see Alessandro Langiu and Maurizio Portaluri, *Di fabbrica si muore* (San Cesario di Lecce: Manni, 2008).
23. See for example: Myrna Santiago, 'Work, Home and Natural Environments: Health and Safety in the Mexican Oil Industry', in *Dangerous Trades. Histories of Industrial Hazards in a Globalizing World*, ed. Christopher Sellers and Joseph Melling (Philadelphia: Temple University Press, 2012), 33–46. See also Christopher Sellers, 'Thoureau's Body: Towards an Embodied Environmental History', *Environmental History* 4 (1999): 486–514; on labour and workers' bodies in environmental history, see also Arthur McEvoy, 'Working Environments: An Ecological Approach to Industrial Health and Safety', *Technology and Culture* 36 (1995): s145–s173; Linda Nash, 'The Fruits of Ill-Health: Pesticides and Workers' Bodies in Post-World War II California', *Osiris* 19 (2004): 203–19; Chad Montrie, *Making a Living: Work and Environment in the United States* (Chapel Hill: University of North Carolina Press, 2008).
24. See McEvoy, 'Working Environments'.

3. REFUSING 'NUCLEAR HOUSEWORK'

1. See Ryan Holifield, Jayajit Chakraborty, and Gordon Walker, *The Routledge Handbook of Environmental Justice* (London: Routledge, 2017); Joan Martínez-Alier, *The Environmentalism of the Poor: A Study of Ecological Conflicts*

and Valuation (Cheltenham: Edward Elgar Publishing, 2002); Elia Apostolopoulou and Jose A. Cortes-Vazquez, *The Right to Nature: Social Movements, Environmental Justice and Neoliberal Natures* (Routledge, 2019); Wendy Harcourt, Ana Agostino, Rebecca Elmhirst, Marlene Gómez, and Panagiota Kotsila, eds, *Contours of Feminist Political Ecology* (Gender, Development and Social Change) (Cham: Springer International Publishing, 2023), https://doi.org/10.1007/978-3-031-20928-4.

2. Great Accleration is a term used in climate science to indicate the period after 1950, when all indicators of earth-system change have risen exponentially. For more details, see the Introduction to this book.
3. Their objection and the cross-examination are integrally published in the short booklet *Refusing Nuclear Housework. Women's Case against the Building of Hinkley 'C' Nuclear Power Station. Testimony to the Hinkley C Public Inquiry from: Wages for Housework Campaign, Black Women for WFH, Win-visible: Women in Visible and Invisible Disabilities and Bristol Women's Peace Collective* (London: Crossroads, 1989).
4. Jane Roberts, 'Clarity, Ambivalence or Confusion? An Assessment of Pressure Group Motives at the Hinkley Point 'C' Public Inquiry', *Energy & Environment* 2, no. 1 (1 March 1991), https://doi.org/10.1177/0958305X9100200103.
5. A major case in point is Italy's anti-nuclear referendum, held in 1987, which ruled out the nuclear option altogether (see Chapter 6). On connections between anti-nuclear and environmental mobilisations in Western Europe, see also Stefania Barca and Ana Delicado, 'Anti-Nuclear Mobilisation and Environmentalism in Europe: A View from Portugal (1976–1986)', *Environment and History* 22, no. 4 (November 2016): 497–520, https://doi.org/10.3197/096734016X14727286515736.
6. For the use of this concept in environmental justice research, see Robert D. Bullard, 'Differential Vulnerabilities: Environmental and Economic Inequality and Government Response to Unnatural Disasters', *Social Research: An International Quarterly* 75, no. 3 (2008), https://doi.org/10.1353/sor.2008.0035.
7. Roberts, 'Clarity, Ambivalence or Confusion?'
8. *Refusing Nuclear Housework*, 32.
9. Carolyn Merchant, *Earthcare: Women and the Environment* (London: Routledge, 1996); Carolyn Merchant, *Radical Ecology: The Search for a Livable World* (London: Routledge, 2005).
10. Rosalie Bertell, *No Immediate Danger: Prognosis for a Radioactive Earth* (London: Women's Press, 1985); 'Rosalie Bertell', Right Livelihood, accessed 13 September 2023, https://rightlivelihood.org/the-change-makers/find-a-laureate/rosalie-bertell/.
11. Stephanie Leland, *Reclaim the Earth: Women Speak out for Life on Earth* (London: Women's Press, 1983).
12. *Refusing Nuclear Housework*, 4; 'Report of the World Conference to Review and Appraise the Achievements of the United Nations Decade for Women: Equality, Development and Peace, Nairobi, 15–26 July 1985' (Nairobi: United Nations, 1986), https://digitallibrary.un.org/record/113822.
13. Roberts, 'Clarity, Ambivalence or Confusion?'

14. Philip Johnstone, 'From Inquiry to Consultation: Contested Spaces of Public Engagement with Nuclear Power' (PhD thesis, University of Exeter, 2013), https://ore.exeter.ac.uk/repository/handle/10871/14255. I wish to thank Simon Pirani and Philip Johnstone for their help in identifying sources about the Hinkley C controversy.
15. Anonymous, interview by Philip Johnstone, 22 March 1989, quoted in ibid., 147.
16. Johnstone, 'From Inquiry to Consultation', 146.
17. *Refusing Nuclear Housework*, 21.
18. I reproduce here the expression originally used by the WFH movement – which was the one widely adopted at the time to refer to what later became 'developing countries', and today is more typically named the 'Global South'.
19. Ibid., 21.
20. Ibid., 4–5.
21. Ibid., 6–8.
22. Ibid., 8–9.
23. Maril Hazlett, '"Woman vs. Man vs. Bugs": Gender and Popular Ecology in Early Reactions to Silent Spring', *Environmental History* 9, no. 4 (October 2004): 701–29, https://doi.org/10.2307/3986266.
24. 'Greenham Common Women's Peace Camp', Wikipedia, 19 July 2023, https://en.wikipedia.org/w/index.php?title=Greenham_Common_Women%27s_Peace_Camp&oldid=1166061560. *Mothers of the Revolution*, Documentary (General Film Corporation, 2021).
25. *Refusing Nuclear Housework*, 8–10.
26. Ibid., 13.
27. Ibid., 13.
28. Glasman cites a study by the West Berlin Human Genetic Institute, quoted in an article on Women's World (June 1987) which found that nine months after the Chernobyl accident the number of babies born with genetic malformations was five times the average (ibid., 14). This observation appears to be confirmed, or even underestimated, by subsequent studies on genetic effects of the accident over the long term. See for example: Anton V. Korsakov, Emilia V. Geger, Dmitry G. Lagerev, Leonid I. Pugach, and Timothy A. Mousseau, 'De Novo Congenital Malformation Frequencies in Children from the Bryansk Region Following the Chernobyl Disaster (2000–2017)', *Heliyon* 6, no. (2020).
29. Ibid., 14.
30. Wilmette Brown, *Roots: Black Ghetto Ecology* (London: King's Cross Women's Centre, 1986).
31. *Refusing Nuclear Housework*, 15.
32. Ibid., 2.
33. Ibid., 17.
34. Ibid., 15–18.
35. Ibid., 18.
36. Ibid., 18.
37. See for example: U.T. Srinivasan, S.P. Carey, E. Hallstein, P.A. Higgins, A.C. Kerr, L.E. Koteen, and R.B. Norgaard, 'The debt of nations and the distribu-

tion of ecological impacts from human activities'. *Proceedings of the National Academy of Sciences* 105 no. 5 (2008): 1768–73; see also Andrew Simms, *Ecological Debt: The Health of the Planet and the Wealth of Nations* (London: Pluto Press, 2005).

38. According to the British Anti Apartheid Movement (AAM), 'In the 1970s and 1980s Britain imported uranium from RTZ's Rossing mine in Namibia in contravention of UN resolutions that said the country's natural resources should only be sold with the consent of the UN Council for Namibia. The uranium was imported under contracts signed in the late 1960s by the UK Atomic Energy Authority and Rio Tinto Zinc. The Campaign Against the Namibian Uranium Contract (CANUC) was set up in 1977 by the Namibia Support Committee, the Haslemere Group and the AAM.' See the AAM online archive: www.aamarchives.org/archive/history/namibia/p0076-cancel-the-namibian-uranium-contract.html, accessed 30 January 2024. More information can be found at: www.ejolt.org/2014/05/watch-a-lamca-ejolt-movie-on-uranium-mining-in-namibia/, accessed 30 January 2024. See also M. Conde and G. Kallis, 'The Global Uranium Rush and Its Africa Frontier. Effects, Reactions and Social Movements in Namibia, *Global Environmental Change* 22, no. 3 (2012): 596–610.
39. *Refusing Nuclear Housework*, 19–21. See also Zoe De Ishtar, 'Nuclearised bodies and militarised space', in *Eco-Sufficiency and Global Justice: Women Write Political Ecology*, ed. Ariel Salleh (London: Pluto Press, 2009), 121–39.
40. Ibid., 22.
41. Ibid., 3.
42. Martínez-Alier, *The Environmentalism of the Poor.*
43. *Refusing Nuclear Housework*, 22.
44. Holifield, Chakraborty, and Walker, *The Routledge Handbook of Environmental Justice.*
45. *Refusing Nuclear Housework*, 22.

4. TAKING CARE OF THE AMAZON

1. Felipe Milanez, 'Ousadia e luta: o pensamento de defensores da floresta na Amazônia', *Boletim do Museu Paraense Emílio Goeldi. Ciências Humanas* 18, no. 2 (2023), https://doi.org/10.1590/2178-2547-bgoeldi-2022-0037; Felipe Milanez, '"A ousadia de conviver com a floresta": uma ecologia política do extrativismo na Amazônia' (PhD thesis, Coimbra, University of Coimbra, 2015), https://estudogeral.sib.uc.pt/handle/10316/29762.
2. 'Defenders of the Earth: Global Killings of Land and Environmental Defenders in 2016' (London: Global Witness, 2017), www.globalwitness.org/en/campaigns/environmental-activists/defenders-earth/.
3. Joan Martínez-Alier, Leah Temper, Daniela Del Bene, and Arnim Scheidel, 'Is There a Global Environmental Justice Movement?' *The Journal of Peasant Studies* 43, no. 3 (3 May 2016), https://doi.org/10.1080/03066150.2016.1141198; Leah Temper, Daniela Del Bene, and Joan Martinez-Alier, 'Mapping the Frontiers and Front Lines of Global Environmental Justice:

The EJAtlas', *Journal of Political Ecology* 22, no. 1 (1 December 2015), https://doi.org/10.2458/v22i1.21108; Leah Temper, Federico Demaria, Arnim Scheidel, Daniela Del Bene, and Joan Martinez-Alier, 'The Global Environmental Justice Atlas (EJAtlas): Ecological Distribution Conflicts as Forces for Sustainability', *Sustainability Science* 13, no. 3 (May 2018), https://doi.org/10.1007/s11625-018-0563-4; Arnim Scheidel et al., 'Environmental Conflicts and Defenders: A Global Overview', *Global Environmental Change* 63 (July 2020), https://doi.org/10.1016/j.gloenvcha.2020.102104.

4. Joan Martínez-Alier, *The Environmentalism of the Poor: A Study of Ecological Conflicts and Valuation* (Cheltenham: Edward Elgar, 2002).
5. Stefania Barca, *Forces of Reproduction: Notes for a Counter-Hegemonic Anthropocene*, 1st edn (Cambridge University Press, 2020), https://doi.org/10.1017/9781108878371.
6. Miriam Lang and Dunia Mokrani, eds, *Más Allá Del Desarrollo* (Quito: Fundación Rosa Luxemburg, Abya-Yala, 2011).
7. David Bollier and Silke Helfrich, *The Wealth of the Commons: A World beyond Market and State* (Amherst: Levellers Press, 2014); George Caffentzis and Silvia Federici, 'Commons against and beyond Capitalism', *Community Development Journal* 49, no. suppl 1 (1 January 2014), https://doi.org/10.1093/cdj/bsu006.
8. Carlos Benítez Trinidad, Stefania Barca, and Felipe Milanez, 'El común y la violencia política en la Amazonía brasileña: la lucha de la Aliança dos Povos da Floresta', *Studia Historica. Historia Contemporánea* 40 (2022), https://doi.org/10.14201/shhc20183689112; Stefania Barca and Felipe Milanez, 'Labouring the Commons: Amazonia's "Extractive Reserves" and the Legacy of Chico Mendes', in *The Palgrave Handbook of Environmental Labour Studies*, ed. Nora Räthzel, Dimitris Stevis, and David Uzzell (Cham: Springer International Publishing, 2021), https://doi.org/10.1007/978-3-030-71909-8_14.
9. Neera M. Singh, 'The Affective Labor of Growing Forests and the Becoming of Environmental Subjects: Rethinking Environmentality in Odisha, India', *Geoforum* 47 (June 2013), https://doi.org/10.1016/j.geoforum.2013.01.010; Gustavo A. García-López, Ursula Lang, and Neera Singh, 'Commons, Commoning and Co-becoming: Nurturing Life-in-Common and Post-Capitalist Futures (An Introduction to the Theme Issue)', *Environment and Planning E: Nature and Space* 4, no. 4 (1 December 2021), https://doi.org/10.1177/25148486211051081.
10. Felipe Milanez and Bernardo Loyola, *Toxic: Amazon* (Vice, 2011). The entire documentary can be found at: www.youtube.com/watch?v=7TLjazh2vWU.
11. Benítez Trinidad, Barca, and Milanez, 'El común y la violencia política en la Amazonía brasileña'.
12. Mauro Barbosa de Almeida, 'A Enciclopédia Da Floresta e a Florestania' (2008), https://mwba.files.wordpress.com/2010/06/2008-almeida-a-florestania-e-a-enciclopedia-da-floresta-texto.pdf.
13. Felipe Milanez, 'Countering the Order of Progress: Colonialism, Extractivism and Re-existence in the Brazilian Amazon', in *Towards a Political Economy of Degrowth*, ed. Ekaterina Chertkovskaya, Alexander Paulsson, and Stefania Barca (London: Rowman & Littlefield International, 2019), 121–36.

14. Ibid.
15. Mauro Barbosa De Almeida, Mary Helena Allegretti, and Augusto Postigo, 'O legado de Chico Mendes: êxitos e entraves das Reservas Extrativistas', *Desenvolvimento e Meio Ambiente* 48 (30 November 2018), https://doi.org/10.5380/dma.v48i0.60499.
16. Kyle Pows Whyte, 'Our Ancestors' Dystopia Now: Indigenous Conservation and the Anthropocene', in *The Routledge Companion to the Environmental Humanities*, ed. Ursula K. Heise, Jon Christensen, and Michelle Niemann (London: Routledge, 2017), 206–15; Winona LaDuke, *All Our Relations: Native Struggles for Land and Life* (Cambridge, MA: South End Press, 1999).
17. Eduardo Viveiros de Castro, 'Alguma coisa vai ter que acontecer', in *Ailton Krenak*, ed. Sérgio Cohn, Encontros (Rio de Janeiro: Azougue, 2015), 5–19.
18. Sérgio Cohn, ed., *Ailton Krenak*, Encontros (Rio de Janeiro: Azougue, 2015); Susanna B. Hecht and Alexander Cockburn, *The Fate of the Forest: Developers, Destroyers, and Defenders of the Amazon* (Chicago: University of Chicago Press, 2010).
19. Ailton Krenak, *Ideias para adiar o fim do mundo* (São Paulo: Companhia das Letras, 2019), 16 (my translation); English edition: *Ideas to Postpone the End of the World* (Toronto: Anansi International, 2020).
20. Krenak, *Ideias para adiar o fim do mundo*, 22.
21. Ailton Krenak, quoted in Nilo Diniz, *Chico Mendes: um grito no ouvido do mundo: como a imprensa cobriu a luta dos seringueiros* (Appris Editora, 2019), 83 (my translation).
22. Barbosa de Almeida, 'A Enciclopédia Da Floresta e a Florestania'.
23. Milanez, 'A ousadia de conviver com a floresta', 63–4.
24. On 'companion species', see Donna Haraway, *Staying with the Trouble: Making Kin in the Chthulucene* (Experimental Futures: Technological Lives, Scientific Arts, Anthropological Voices) (Durham: Duke University Press, 2016): 9–29; on 'kindred being', see V. Plumwood, 'Nature in the Active Voice', *Australian Humanities Review* 46 (2009): 113–29.
25. On the concept of 'death of nature', see Carolyn Merchant, *The Death of Nature: Women, Ecology, and the Scientific Revolution* (New York: Harper & Row, 1989).
26. Maristella Svampa, *Neo-extractivism in Latin America: Socio-Environmental Conflicts, the Territorial Turn, and New Political Narratives* (New York: Cambridge University Press, 2019).
27. I use this concept as distinct from the broader concept of anthropocentrism, which I find less useful and constructive in political terms.
28. Merchant, *The Death of Nature*.
29. Teresa Brennan, *Exhausting Modernity: Grounds for a New Economy* (London: Routledge, 2000).
30. Val Plumwood, *Environmental Culture: The Ecological Crisis of Reason* (Routledge, 2005), 176.
31. Stacy Alaimo, 'Your Shell on Acid: Material Immersion, Anthropocene Dissolves', in *Anthropocene Feminism*, ed. Richard Grusin (Minneapolis: University of Minnesota Press, 2017), 89–120; Stacy Alaimo, *Bodily Natures:*

Science, Environment, and the Material Self (Bloomington: Indiana University Press, 2010).

32. Haraway, *Staying with the Trouble: Making Kin in the Chthulucene*, 100.
33. Ibid.
34. Greta Gaard, 'Ecofeminism Revisited: Rejecting Essentialism and Re-placing Species in a Material Feminist Environmentalism', *Feminist Formations* 23, no. 2 (2011): 26–53; Deborah Bird Rose, 'Val Plumwood's Philosophical Animism: Attentive Interactions in the Sentient World', *Environmental Humanities* 3, no. 1 (2013), https://doi.org/10.1215/22011919-3611248; Greta Gaard, ed.,*Ecofeminism: Women, Animals, Nature* (Philadelphia: Temple University Press, 1993).
35. Catriona Mortimer-Sandilands and Bruce Erickson, eds, *Queer Ecologies: Sex, Nature, Politics, Desire* (Bloomington: Indiana University Press, 2010); Greta Gaard, 'Toward a Queer Ecofeminism', *Hypatia* 12, no. 1 (1997), https://doi.org/10.1111/j.1527-2001.1997.tb00174.x.
36. Gerda Roelvink, 'Rethinking Species-Being in the Anthropocene', *Rethinking Marxism* 25, no. 1 (January 2013): 59, https://doi.org/10.1080/08935696.2012.654700. For a Marxist appraisal of plants' work, see also Marion Ernwein, Franklin Ginn, and James Palmer, eds, *The Work That Plants Do. Life, Labour and the Future of Vegetal Economies* (Transcript: Verlag, 2021).
37. Alyssa Battistoni, 'Bringing in the Work of Nature: From Natural Capital to Hybrid Labor', *Political Theory* 45, no. 1 (February 2017): 5, https://doi.org/10.1177/0090591716638389. On the work of plants see also Ernwein et al., *The Work That Plants Do.*
38. Ibid., 6.

5. GREENING THE JOB

1. Stefania Barca, 'On Working-Class Environmentalism: A Historical and Transnational Overview', *Interface: A Journal for and about Social Movements*, 4, no. 2 (2012): 61–80.
2. Raymond Williams, 'Ideas of Nature', in *Problems in Materialism and Culture*, ed. Raymond Williams (London: Verso, 1980), 67–85.
3. Stefania Barca, 'Laboring the Earth: Transnational Reflections on the Environmental History of Work', *Environmental History* 19 (2014): 3–27.
4. Eric Bonds and Liam Downey, '"Green" Technology and Ecologically Unequal Exchange: The Environmental and Social Consequences of Ecological Modernization in the World-System', *Journal of World System Research* 18 (2012): 167–86; Adam G. Bumpus and Diana M. Liverman, 'Carbon Colonialism? Offsets, Greenhouse Gas Reductions, and Sustainable Development', in *Global Political Ecology*, ed. Richard Peet, Peter Robbins, and Michael Watts (New York: Routledge, 2011), 203–24; Noel Castree, 'Crisis, Continuity and Change: Neoliberalism, the Left and the Future of Capitalism', *Antipode* 41 (2009): 185–213; Les Levidow and Helena R. Paul, 'Global Agrofuel Crops as Contested Sustainability, Part II: Eco-efficient Techno-fixes?' *Capitalism Nature Socialism* 22, no. 2 (2011): 27–51.

5. See David Schwartzman, 'Green New Deal: An Ecosocialist Perspective', *Capitalism Nature Socialism* 22, no. 3 (2011): 49–56 (49). For a feminist appraisal, see Tithi Batthacharya, 'Three Ways a Green New Deal Can Promote Life over Capital'. *Jacobin* (6 October 2019), accessed 30 January 2024, https://jacobinmag.com/2019/06/green-new-deal-social-care-work. For a decolonial appraisal, see Max Ajl, *A People's Green New Deal* (London: Pluto Press, 2021).
6. See Don Fitz, 'How Green Is the Green New Deal?' *Climate and Capitalism* 7, accessed 30 January 2024, http://climateandcapitalism.com/2014/07/15/green-green-new-deal.
7. Ibid.
8. Ibid.
9. Jesse Goldstein, 'Appropriate Technocracies? Green Capitalist Discourses and Post Capitalist Desires', *Capitalism Nature Socialism* 24 (2013): 26–34.
10. Sara Nelson, 'Rio+20, Climate Change, and Critical Scholarship: Beyond the Critique of "green neoliberalism"?', entry in Antipode Foundation.org, accessed 15 September 2014, http://wp.me/p16RPC-re.
11. See Boone Shear, 'Making the Green Economy: Politics, Desire, and Economic Possibility', *Journal of Political Ecology* 21 (2014): 193–209 (195).
12. Ibid., 197.
13. Ibid., 201.
14. Ibid., 205.
15. Ibid., 206.
16. See Nora Räthzel and David Uzzell, 'Mending the Breach between Labor and Nature: A Case for Environmental Labor Studies', in *Trade Unions in the Green Economy*, ed. Nora Räthzel and David Uzzell (London: Routledge, 2013), 1–12.
17. Ibid., 9.
18. Ibid., 10.
19. Ibid., 11.
20. See *The Palgrave Handbook of Environmental Labour Studies*, ed. Nora Räthzel, Dimitris Stevis, and David Uzzell (Cham: Palgrave Macmillan, 2021).
21. Jacklyn Cock, 'The "Green Economy": A Just and Sustainable Development Path or a "Wolf in Sheep's Clothing"?', *Global Labour Journal* 5, no. 1 (31 January 2014): 23–44, https://doi.org/10.15173/glj.v5i1.1146.
22. Ibid., 24.
23. Ibid., 39.
24. See Patricia E. Perkins, 'Feminist Ecological Economics', in *Economics Interactions with Other Disciplines*, vol. 2, *Encyclopedia of Life Support Systems* (EOLSS), ed. M. Gowdy (Paris: EOLSS Publishers, 2009), 1, www.eolss.net.
25. For example, see Maria Mies and Veronika Bennholdt-Thomsen, *The Subsistence Perspective. Beyond the Globalised Economy* (1997; reprinted London and New York: Zed Books, 1999).
26. See Ariel Salleh, 'From Metabolic Rift to "Metabolic Value": Reflections on Environmental Sociology and the Alternative Globalization Movement',

Organization & Environment 23, no. 2 (1 June 2010): 205–19, https://doi.org/10.1177/1086026610372134.

27. Ibid.
28. Ibid.
29. See J.K. Gibson-Graham, 'Diverse Economies: Performative Practices for "Other Worlds"', *Progress in Human Geography* 32 (2008): 613–32 (615).
30. Ibid.
31. Ibid., 617.
32. See J.K. Gibson-Graham, *The End of Capitalism (As We Knew It)* (Minneapolis: University of Minnesota Press, 2006), ix.
33. See J.K. Gibson-Graham and Gerda. Roelvink (2010), 'An Economic Ethics for the Anthropocene', *Antipode* 41 (2010): 320–46 (342).
34. See Brian J. Burke and Boone Shear, 'Introduction: Engaged Scholarship for Non-Capitalist Political Ecologies', *Journal of Political Ecology* 21(2014): 127–44 (129).
35. See Silvia Federici, *Revolution at Point Zero* (Oakland: PM Press, 2012);see Chapter 3 in this book.
36. See Christine Bauhardt, 'Solutions to the Crisis? The Green New Deal, Degrowth, and the Solidarity Economy: Alternatives to the Capitalist Growth Economy from an Ecofeminist Economics Perspective', *Ecological Economics* 102 (1 June 2014): 60–8, https://doi.org/10.1016/j.ecolecon.2014.03.015.
37. See Barca, 'Laboring the Earth'. M. Michael Mason and Nigel Morter, 'Trade Unions as Environmental Actors: The UK Transport and General Workers' Union', *Capitalism Nature Socialism* 9, no. 2 (1998): 3–34; Uzzell and Räthzel, 'Mending the Breach between Labour and Nature'.
38. See, for example, Dimitris Stevis, 'Just Transitions: Promise and Contestation', in *Elements in Earth System Governance* (April 2023), https://doi.org/10.1017/9781108936569.
39. See also the Epilogue of this book.
40. Barca, 'On working-Class Environmentalism'.
41. For further elaboration of this point, see Irina Velicu and Stefania Barca, 'The Just Transition and Its Work of iIequality', *Sustainability: Science, Practice and Policy* 16, no. 1, 263–73.
42. K.A. Kenneth Gould, Tammy L. Lewis and Timmons Robins, 'Blue–Green Coalitions: Constraints and Possibilities in the Post 9-11 Political Environment', *Journal of World System Research* 10, no.1 (2004): 91–116.
43. See Anabella Rosenberg, 'Building a Just Transition: The Linkages between Climate Change and Employment', *International Journal of Labour Research* 2 (2010): 125–62 (141).
44. Ibid., 141.
45. Ibid., 144.
46. See Dan Cunniah, 'Preface, Climate Change and Labour: The Need for a "Just Transition"', *International Journal of Labour Research* 2 (2010): 121–3 (121).
47. Rosenberg, 'Building a Just Transition'.
48. Nicholas Stern, *Stern Review on the Economics of Climate Change* (London: HM Treasury, 2006).

49. See Rosenberg, 'Building a Just Transition', 129.
50. Ibid., 134.
51. Ibid., 139.
52. Ibid., 137.
53. Paulo Sergio Muçouçah, *Empregos Verdes no Brasil: Quantos São, Onde Estão e Como Evoluirão nos Próximos Anos* [Green Jobs in Brazil: How Many There Are, Where, and How Are They Going to Evolve in the Near Future] (São Paulo: Organização Internacional do Trabalho, Escritório do Brasil, 2009).
54. Gilberto Penha de Araújo, 'Qual é o efeito da globalização na exploração da mão de obra no setor sucro alcooleiro no Brasil?' [What Is the Effect of Globalization in the Exploitation of Workers in Sugar-Alcohol Production in Brazil?] (MA thesis, Lisbon New University, 2011).
55. Marcelo Firpo de Souza Porto, Renan Finamore, and Hugo Ferreira, 'Injustiças da sustentabilidade: conflitos ambientais relacionados à produção de energia "limpa" no Brasil' [The Injustice of Sustainability: Environmental Conflicts Related to "Clean" Energy Production in Brazil], *Revista Crítica de Ciências Sociais* [*Critical Journal of Social Sciences*], 100 (2013): 37–64.
56. Muçouçah, *Empregos verdes no Brasil.*
57. See Lene Olsen, 'Supporting a Just Transition: The Role of International Labour Standards', *International Journal of Labour Research* 2 (2010): 293–318 (297).
58. Ibid.
59. Cock, 'The "Green Economy"', 24.
60. Ibid., 24.
61. Ibid., 30.
62. Ibid., 32.
63. Ibid., 32.
64. Ibid., 32.
65. Ibid., 36.
66. 'Climate Jobs Booklet 2011–2', OMCJ [One Million Climate Jobs campaign], accessed 30 September 2014, www.climatejobs.org.za/index.php/research/campaign-booklets.
67. Cock, 'The "Green Economy"', 40.
68. Jacklyn Cock, 'Labour's Response to Climate Change', *Global Labour Journal* 2 (2011): 235–42.
69. See Emmanuele Leonardi, 'Biopolitics of Climate Change: Carbon Commodities, Environmental Profanations, and the Lost Innocence of Use Value' (PhD thesis, University of Western Ontario, 2012), 312.
70. Ibid., 312.
71. Ibid., 314.
72. See OMCJ, 'Climate Jobs Booklet 2011–2'.

6. LABOUR AND THE ECOLOGICAL CRISIS

1. Geoff Eley, *Forging Democracy: The History of the Left in Europe, 1850–2000* (Oxford: Oxford University Press, 2002); Geoff Eley and Keith Nield, 'Farewell to the Working Class?' *International Labor and Working-Class*

History 57 (2000): 1–30; Beverly J. Silver, *Forces of Labor. Workers' Movements and Globalization since 1870* (Cambridge: Cambridge University Press, 2003); Marcel Van der Linden, *Workers of the World: Essays toward a Global Labor History* (Leiden and Boston: Brill, 2008).

2. Stefania Barca, 'Laboring the Earth: Transnational Reflections on the Environmental History of Work', *Environmental History* 19, no. 1 (2014): 3–27; Martin Ryle and Kate Soper, 'Introduction: The Ecology of Labour', *Green Letters* 20, no. 2 (2016): 119–26.
3. In his landmark *Marx's Ecology* (2000), J.B. Foster recalled how – according to Marx – industrial capitalism has turned social metabolism into a 'metabolic rift', that is, a process of accelerated degradation of both labour and non-human nature. See John Bellamy Foster, *Marx's Ecology. Materialism and Nature* (New York: Monthly Review Press, 2000). See also John Bellamy Foster, Brett Clark, and Richard York, *The Ecological Rift: Capitalism's War on the Earth* (New York: Monthly Review Press, 2010); Brett Clark and Richard York, 'Carbon Metabolism: Global Capitalism, Climate Change, and the Biospheric Rift', *Theory and Society* 34, no. 4 (2005): 391–428.
4. Ariel Salleh, 'From Metabolic Rift to "Metabolic Value": Reflections on Environmental Sociology and the Alternative Globalization Movement', *Organization & Environment* 23, no. 2 (2010): 205–19.
5. Ibid., 205.
6. See Ariel Salleh, 'Green Economy' or Green Utopia: The Salience of Reproductive Labor Post-Rio+20', *Journal of World-Systems Research* 18, no. 2 (2012): 138–45 (141).
7. Christine Bauhardt, 'Solutions to the Crisis? The Green New Deal, Degrowth, and the Solidarity Economy: Alternatives to the Capitalist Growth Economy from an Ecofeminist Economics Perspective', *Ecological Economics* 102 (2014): 60–8; M. Mellor, 'Ecofeminist Political Economy', *International Journal of Green Economics* 1, no. 1/2 (2006): 139–50; Patricia E. Perkins, 'Feminist Ecological Economics and Sustainability', *Journal of Bioeconomics* 9 (2007): 227–44.
8. Ariel Salleh, 'CNS Symposium: Ecofeminist Dialogues', *Capitalism Nature Socialism* 17, no. 4 (2006): 32–141.
9. Ramachandra Guha and Joan Martínez Alier, *Varieties of Environmentalism: Essays North and South* (London: Earthscan, 1997); David Harvey, *Justice, Nature and the Geography of Difference* (Oxford and Malden: Blackwell, 1996).
10. Damian White, Alan Rudy, and Brian Gareau, *Environments, Natures and Social Theory: Towards a Critical Hybridity* (London: Macmillan Education UK, 2016).
11. T.W. Luke, 'Environmentality', in *The Oxford Handbook of Climate Change and Society*, ed. John S Dryzek, Richard B. Norgaard, and David Schlosberg (Oxford: Oxford University Press, 2011), 97–109.
12. See Maria Kaika, '"Don't Call Me Resilient Again!": The New Urban Agenda as Immunology. Ellipsis or Ellipsis, What Happens When Communities Refuse to Be Vaccinated with "Smart Cities" and Indicators', *Environment and Urbanization*, 29, no. 1 (2017): 89–102 (97).

13. Ibid., 91.
14. James E. Goodman and Ariel Salleh, 'The "Green Economy": Class Hegemony and Counter-hegemony', *Globalizations* 10, no. 3 (2014): 411–24.
15. Or else, the 'environmentalism of the poor': see Joan Martínez Alier, *The Environmentalism of the Poor: A Study of Ecological Conflicts and Valuation* (Cheltenham: Edward Elgar, 2002).
16. Jacklyn Cock, 'The "Green Economy": A Just and Sustainable Development Path or a "Wolf in Sheep's Clothing"?' *Global Labour Journal* 5, no. 1 (2014): 23–44; Romain Felli, 'An Alternative Socio-ecological Strategy? International Trade Unions' Engagement with Climate Change', *Review of International Political Economy* 21, no. 2 (2014): 372–98; Nora Räthzel and David Uzzell, 'Mending the Breach between Labor and Nature: A Case for Environmental Labor Studies', in *Trade Unions in the Green Economy*, ed. Nora Räthzel and David Uzzell (London: Routledge, 2013), 1–12.
17. Verity Burgmann, 'From "Jobs versus Environment" to "Green-Collar Jobs": Australian Trade Unions and the Climate Change Debate', in Räthzel and Uzzell, eds, *Trade Unions in the Green Economy*, 131–45; Darryn Snell and Peter Fairbrother, 'Just Transition and Labour Environmentalism in Australia', in *Trade Unions in the Green Economy*, 146–61; Dimitris Stevis, 'Green Jobs? Good Jobs? Just Jobs? US Labour Unions Confront Climate Change', in *Trade Unions in the Green Economy*, 179–95; Sean Sweeney, 'US Trade Unions and the Challenge of "Extreme Energy": The Case of the TransCanada Keystone XL Pipeline', in *Trade Unions in the Green Economy*, 196–213.
18. Meg Gingrich, 'From Blue to Green: A Comparative Study of Blue-Collar Unions' Reactions to the Climate Change Threat in the United States and Sweden', in Räthzel and Uzzell, eds, *Trade Unions in the Green Economy*, 214–26; Laura Martín Murillo, 'From Sustainable Development to a Green and Fair Economy: Making the Environment a Trade Union Issue', in *Trade Unions in the Green Economy*, 29–40.
19. Begoña María Tomé-Gil, 'Moving towards Eco-unionism: Reflecting the Spanish Experience', in Räthzel and Uzzell, eds, *Trade Unions in the Green Economy*, 64–77.
20. 'ETUC Action Programme 2015–2019', ETUC (European Trade Unions Confederation, last modified 29 September 2015, www.etuc.org/en/publication/etuc-action-programme-2015-2019.
21. 'Global Climate Jobs', One Million Climate Jobs (OMCJ), accessed 20 May 2018, https://globalclimatejobs.wordpress.com/.
22. Paul Burkett, *Marx and Nature: A Red and Green Perspective* (New York St. Martin's Press, 1999); Brett Clark, 'Marx's Natures: A Response to Foster and Burkett', *Organization & Environment* 14, no. 4 (2001): 432–42; Foster, *Marx's Ecology*; Jason W. Moore, 'Ecology, Capital, and the Nature of Our Times: Accumulation & Crisis in the Capitalist World-Ecology', *Journal of World-Systems Research* XVII, no. 1 (2011):108–47.
23. The difference between the two visions lies in the value attributed to work: meaningless but necessary toil in the first case, potentially creative activity in the second. Clearly, such distinction can bear only in abstract terms, whereas in the historical experience of human beings the two forms of work coexist

and complement each other; in fact, one could not exist without the other. So are the two forms of struggle: for those activities that are mainly painful or repetitive, and cannot be eliminated even in a disalienated social system, democratic control over technology and organisation will be needed in order to reduce them to a minimum, to be carried out in the best possible conditions. All the remaining activities, in which creativity and realisation of human potential can be achieved, will need to be socialised as much as possible, so that they do not remain a prerogative of certain social classes, and of one gender only. See John Bellamy Foster, 'The Meaning of Work in a Sustainable Society: A Marxian View', CUSP. Essay Series on the Morality of Sustainable Prosperity no. 3, last modified 17 March 2017, www.cusp.ac.uk/essay/m1-3/.

24. Bellamy Foster and Brett Clark, 'Marx's Ecology and the Left', *Monthly Review* 68, last modified 1 June 2016, https://monthlyreview.org/2016/06/01/marxs-ecology-and-the-left/.
25. Giovanni was the younger brother of Enrico Berlinguer, who became PCI's secretary general in 1972. A physician by training, he was the author of numerous works on the history of medicine, and became a reference for Marxist occupational health science in and beyond Italy.
26. Wilko Graf von Hardenberg and Paolo Pelizzari, 'The Environmental Question, Employment and Development in Italy's Left', *Left History* 2 (2008): 77–104.
27. Stefania Barca, 'Work, Bodies, Militancy: The "Class Ecology" Debate in 1970s Italy', in *Powerless Science? Science and Politics in a Toxic World*, ed. Soraya Boudia and Nathalie Jas, vol. 2, The Environment in History: International Perspectives (New York: Berghahn Books, 2014), 115–33. See also Chapter 1 in this book.
28. Laura Conti, *Che cos'è l'ecologia. Capitale, lavoro, ambiente* [What Is Ecology. Capital, Labour, Environment] (Milan: Mazzotta, 1977), 140n.
29. Ibid., 136–9.
30. See Chapter 2 in this book.
31. Among them, the urban ecologist Virginio Bettini, who had co-authored with Barry Commoner a book called *Ecologia e lotte sociali. Ambiente, popolazione, inquinamento* [Ecology and Social Struggles. Environment, Population, Pollution] (1976), putting forth the thesis of the two ecologies (ecology of power and ecology of class), which became a political manifesto for the Italian Left ecology. Yet another founding member of Legambiente was the chemist Giorgio Nebbia, also a well-known academic and the author of a number of books and pamphlets on political ecology (see Barca, 'Work, Bodies, Militancy').
32. Donatella Della Porta and Mario Diani, *Movimenti senza protesta? L'ambientalismo in Italia* [Movements without Protest? Environmentalism in Italy] (Bologna: Il Mulino, 2004); Roberto Della Seta, *La difesa dell'ambiente in Italia* [The Defence of the Environment in Italy] (Milan: Franco Angeli, 2000).

33. The claim is attributed to Andrea Poggio, journalist and author, one of the founders of Legambiente and director of the magazine *La Nuova Ecologia* between 1980 and 1984; see Della Seta, *La difesa dell'ambiente in Italia*, 50–1.
34. Ibid.
35. The magazine, a sister publication within the international family of Western eco-Marxism, namely, in connection with the US-based *Capitalism Nature Socialism*, the French Écologie *et Politique* and the Spanish *Ecología Política*, was directed by Giovanna Ricoveri and Valentino Parlato (who was also the director of the newspaper *Il Manifesto*).
36. The article postulated that the point of departure of ecological socialism was the contradiction between capitalist forces and relations of production and the 'conditions of production', which he viewed – following Polanyi – as 'fictitious commodities'. This contradiction would cause ecological crisis as a 'crisis of underproduction', that is, a non-Malthusian version of scarcity, whereby capital induces the destruction of the conditions of production. This kind of scarcity, O'Connor believed, would lead to an increased socialisation of production, via economic planning and environmental regulations, thus creating the possibility for an ecological path to socialism. See James O'Connor, 'Capitalism, Nature, Socialism: A Theoretical Introduction', *Capitalism, Nature, Socialism* 1, no. 1 (1988): 11–38.
37. Examples are Mumford's critique of technology, or Sachs' and Latouche's critique of Western development. For a (partial) list of articles published in the 1990s, see *Ecología Política*, accessed 4 March 2018, www.ecologiapolitica.info.
38. André Gorz, 'Ecology and Freedom', in *Ecology as Politics* (Boston: South End Press, 1979) [*Écologie et liberté*, Galilée, 1977].
39. André Gorz, *Critique of Economic Reason* (London: Verso, 1989) [*Métamorphoses du travail*, Galilée, 1988].
40. Emmanuele Leonardi, 'Introduzione', in *Ecologia e libertà*, ed. André Gorz (Napoli-Salerno: Orthotes, 2015), 17.
41. André Gorz, 'Critique of Economic Reason: Summary for Trade Unionists and Other Left Activists', in *Labour Worldwide in the Era of Globalization: Alternative Union Models in the New World Order*, ed. Ronaldo Munck and Peter Waterman (London: Macmillan, 1999), 41–63.
42. André Gorz, *Ecologica* (Salt Lake City: Seagull Books, 2010), 9.
43. Gorz, 'Ecology and Freedom'.
44. André Gorz, *Farewell to the Working Class: An Essay on Post-Industrial Socialism* (London and Sydney: Pluto Press, 1982) [*Adieux au Prolétariat*, Galilée, 1980].
45. Ibid., 66.
46. Gorz, *Ecologica*, 11.
47. Gorz, *Farewell to the Working Class*, 8.
48. Ibid., 14–15.
49. Ibid., 11.
50. Ibid., 12.
51. See Ryle and Soper, 'Introduction: The Ecology of Labour', 120.

52. On reduction of working time and degrowth, see especially chapters 14 and 15 in Giorgos Kallis, *In Defence of Degrowth: Opinions and Manifesto*, ed. Aaron Vansintjan, accessed 13 October 2017, https://indefenseofdegrowth.com.
53. Giacomo D'Alisa, Federico Demaria, and Giorgos Kallis, *Degrowth: A Vocabulary for a New Era* (London and New York: Routledge, 2014).
54. Christopher Rootes, *Environmental Protest in Western Europe* (Oxford: Oxford University Press, 2003).
55. Raymond Williams, 'Ecology & the Labour Movement'. A talk given at the Plinston Hall, Letchworth, 2 June 1984, accessed 31 January 2024, www.youtube.com/watch?v= EiFWHtKOcjo. I wish to thank Jason W. Moore for posting this resource on the World Ecology Research Network, through which I came to know about it.
56. O'Connor, 'Capitalism, Nature, Socialism'.
57. Williams, 'Ecology & the Labour Movement'.
58. Moore, 'Ecology, Capital and the Nature of Our Times'.
59. Cinzia Arruzza, 'Functionalist, Determinist, Reductionist: Social Reproduction Feminism and Its Critics', *Science & Society* 80, no. 1 (2016): 9–30; Mariarosa Dalla Costa, 'Introduction to the Archive of Feminist Struggle for Wages for Housework', *Viewpoint Magazine*, issue 5, October, accessed 31 January 2024, https://viewpointmag.com/2015/11/02/issue-5-social-reproduction/; Nancy Fraser, 'Behind Marx's Hidden Abode: For an Expanded Conception of Capitalism', *New Left Review* 86 (March–April 2014): 55–72.
60. Maria Mies, *Patriarchy and Accumulation on the World Scale* (London: Zed Books, 1996).
61. Ibid., 48.
62. Ibid., 52.
63. Ibid., 53.
64. Ibid., 69.
65. Maria Mies and Vandana Shiva, *Ecofeminism* (London: Zed Books, 1993).
66. Maria Mies and Veronika Bennholdt-Thomsen, *The Subsistence Perspective* (London: Zed Books, 2000).
67. Maria Mies, 'Questioning Needs: A Rejoinder to Victor Wallis', *Capitalism Nature Socialism* 17, no. 4 (2006): 44–7; Silvia Federici, *Revolution at Point Zero. Housework, Reproduction, and Feminist Struggle* (Oakland: PM Press, 2012).
68. Mariarosa Dalla Costa,'The Native in Us, the Earth We Belong to', *The Commoner* 6 (Winter 2003), accessed 10 April 2016, www.thecommoner.org; Federici, *Revolution at Point Zero*; Mary Mellor, *Feminism and Ecology* (New York: New York University Press, 1999).
69. Salleh 'From Metabolic Rift to "Metabolic Value"', 205–19.
70. Salleh, 'CNS Symposyum', 32–141.
71. Kate Soper. 'Feminism and Ecology: Realism and Rhetoric in the Discourses of Nature', *Science, Technology & Human Values* 20, no. 3 (1995): 311–31.
72. Carolyn Merchant, *Radical Ecology: The Search for a Livable World* (London: Routledge, 2005); Ariel Salleh, *Ecofeminism as Politics: Nature, Marx and the Post-Modern* (London: Zed Books, 1997).

73. Greta Gaard, 'Ecofeminism Revisited: Rejecting Essentialism and Replacing Species in a Material Feminist Environmentalism', *Feminist Form* 23, no. 2 (2011): 26–53.
74. See Kallis, *In Defence of Degrowth*, 15.

7. THE LABOUR(S) OF DEGROWTH

1. John Bellamy Foster, 'Marxism and Ecology: Common Fonts of a Great Transition', The great transition initiative, last modified 1 October 2015, www.greattransition.org/ publication/marxism-and-ecology; Giorgos Kallis, 'Is There a Growth Imperative in Capitalism? A Commentary on John Bellamy Foster', Entitle blog, last modified 27 October 2015, https:// entitleblog.org/2015/10/27/is-there-a-growth-imperative-in-capitalism-a-response-to-john-bellamy- foster-part-i/.
2. Lee Brownhill, Teresa E. Turner, and Wahu Kaara, 'Degrowth? How about Some "De-alienation"?' *Capitalism Nature Socialism* 23, no. 1 (2012): 93–104.
3. Ekaterina Chertkovskaya and Alexander Paulsson 'The Growthocene: Thinking through What Degrowth Is Criticising', Entitle blog, last modified 19 February 2016, https:// entitleblog.org/2016/02/19/the-growthocene-thinking-through-what-degrowth-is-criticising/.
4. Stefania Barca, 'Telling the Right Story: Environmental Violence and Liberation Narratives', *Environment and History* 20, no. 4 (2014): 535–46.
5. Giacomo D'Alisa, Federico Demaria, and Giorgos Kallis, *Degrowth: A Vocabulary for a New Era* (London and New York: Routledge, 2014).
6. Brownhill, Turner, and Kaara, 'Degrowth?'
7. André Gorz, *Farewell to the Working Class: An Essay on Post-Industrial Socialism* (London and Sydney: Pluto Press, 1982). See also Chapter 6 in this book.
8. I am grateful to Giacomo D'Alisa for pointing me to the issue of clashing social metabolisms.
9. José Manuel Rojas, '"I Work There, I Know What They Do": The Testimony of an Indigenous Oil Worker in Patagonia', The Leap: System Change on a Deadline, 2016, https://theleapblog.org/i-work-there-i-know-what-they-do-the-testimony-of-an- indigenous-oil-worker-in-patagonia/ (blog no longer available).
10. See Chapter 4 in this book.
11. 'Da Maflow a Rimaflow', Rimaflow. Reditto, Lavoro, Dignitá, Autogestione, accessed 30 May 2016. www.rimaflow.it/index.php/la-nostra-storia/.
12. See also Chapter 5 in this book.
13. Paul Sweeney, 'U.S. Trade Unions and the Challenge of "Extreme Energy"': The Case of the TransCanada Keystone XL Pipeline', in *Trade Unions in the Green Economy*, 196–213.
14. See also Chapter 5 in this book.
15. Denis Bayon, 'Unions', in *Degrowth: A Vocabulary for a New Era*, ed. Giacomo D'Alisa, Federico Demaria, and Giorgos Kallis (London and New York: Routledge, 2014), 189–191.

16. Stefania Barca, 'Labour and Climate Change: Towards an Emancipatory Ecological Class Consciousness', in *Refocusing Resistance to Climate Justice: COPing in, COPing out and beyond Paris*, ed. Leah Temper and Tamra Gilbertson, 74–78, Environmental Justice Organizations, Liabilities and Trade, report no. 23, last modified 1 September 2015, www.ejolt.org/wordpress/wp-content/uploads/2015/09/EJOLT-6.74-78.pdf.
17. Giacomo D'Alisa, Federico Demaria, and Marco Deriu, 'Care', in *Degrowth: A Vocabulary for a New Era*, ed. Giacomo D'Alisa, Federico Demaria, and Giorgos Kallis (London and New York: Routledge, 2014), 96–100.
18. Ariel Salleh, 'From Metabolic Rift to "Metabolic Value": Reflections on Environmental Sociology and the Alternative Globalization Movement', *Organization & Environment* 23, no. 2 (2010): 205–19.
19. Nancy Fraser, 'Behind Marx's Hidden Abode: For an Expanded Conception of Capitalism', *New Left Review* 86 (March–April 2014): 55–72; Jason W. Moore, *Capitalism in the Web of Life: Ecology and the Accumulation of Capital* (London: Routledge, 2015).
20. Christine Bauhardt, 'Solutions to the Crisis? The Green New Deal, Degrowth, and the Solidarity Economy: Alternatives to the Capitalist Growth Economy from an Ecofeminist Economics Perspective', *Ecological Economics* 102 (2014): 60–8.
21. Silvia Federici, *Revolution at Point Zero. Housework, Reproduction, and Feminist Struggle* (Oakland: PM Press, 2012); Carolyn Merchant, *Ecological Revolutions: Nature, Gender and Science in New England* (Chapel Hill: University of North Carolina Press, 2010); Christa Wichterich, 'Contesting Green Growth, Connecting Care, Commons and Enough', in *Practicing Feminist Political Ecologies. Moving beyond the 'Green Economy'*, ed. Wendy Harcourt and Ingrid L. Nelson (London: Zed Books, 2015), 67–100.
22. Wendy Harcourt and Ingrid Nelson, *Practicing Feminist Political Ecologies: Moving beyond the 'Green Economy'* (London: Zed Books, 2015).

8. EPILOGUE

1. At the time of writing, eight country/regional campaigns seem to exist: Norway, United States, Portugal, Canada, UK, France, South Africa and Scotland. See Global Climate Jobs, 'Global Climate Jobs', accessed 29 September 2023, www.globalclimatejobs.org/.
2. These quotes are my translation from the Portuguese climate jobs plan of 2021. See Empregos para o Clima, 'Relatório "Empregos para o Clima" – Empregos para o Clima', accessed 29 September 2023. www.empregos-clima.pt/relatorio/.
3. Two paramount examples in Europe are the projected lithium mining in the Cova do Barroso valley of Northern Portugal, where anti-mining mobilisations are being backed up by climate justice activists nationally and internationally; and the TAV high-speed railway under construction between Turin and Lyonne, vibrantly opposed by local communities of the Susa valley for almost three decades, despite harsh police repression.

4. See Chapters 5, 6 and 7 in this book.
5. The Care Collective et al., *The Care Manifesto: The Politics of Interdependence* (London: Verso, 2020); Laura Addati, Umberto Cattaneo, Valeria Esquivel, and Isabel Valarino, 'Care Work and Care Jobs for the Future of Decent Work', Report, 28 June 2018, accessed 31 January 2024, www.ilo.org/global/publications/books/WCMS_633135/lang--en/index.htm; Stefania Barca, Giacomo D'Alisa, Selma James, and Nina López, *Renta de Los Cuidados ya!* (Barcelona: Icaria; El Viejo Topo; Montaber, 2020), www.icariaeditorial.com/pdf/Renta_de_los_cuidados_definitiu.pdf; Tithi Bhattacharya, *Social Reproduction Theory: Remapping Class, Recentering Oppression* (London: Pluto Press, 2017); Clare Coffey et al., 'Time to Care: Unpaid and Underpaid Care Work and the Global Inequality Crisis' (Oxfam, 20 January 2020), https://doi.org/10.21201/2020.5419.
6. See 'Feminist Green New Deal', accessed 30 January 2024, https://feministgreennewdeal.com/
7. See 'A Blueprint for Europe's Just Transition', Green New Deal for Europe, accessed 30 January 2024, https://report.gndforeurope.com/.
8. 'Care Income Now', Global Women's Strike, accessed 30 January 2024, https://globalwomenstrike.net/care-income-now/. On the GWS and its connection to the Wages for Housework Campaign, see also the Introduction and Chapter 3 in this book.

Index

Thanks to our Patreon subscriber:

Ciaran Kane

Who has shown generosity and comradeship in support of our publishing.

Check out the other perks you get by subscribing to our Patreon – visit patreon.com/plutopress.

Subscriptions start from £3 a month.